THE ALCHEMIST
From a mezzotint by William Pether after Joseph Wright

Alchemy and Alchemists

C. J. S. Thompson

DOVER PUBLICATIONS, INC.
Mineola, New York

Published in Canada by General Publishing Company, Ltd., 895 Don Mills Road, 400-2 Park Centre, Toronto, Ontario M3C 1W3.
Published in the United Kingdom by David & Charles, Brunel House, Forde Close, Newton Abbot, Devon TQ12 4PU.

Bibliographical Note

This Dover edition, first published in 2002, is an unabridged republication of the work published as *The Lure and Romance of Alchemy* by George G. Harrap & Company Ltd., London, in 1932.

Library of Congress Cataloging-in-Publication Data

Thompson, C. J. S. (Charles John Samuel), 1862–1943.
 [Lure and romance of alchemy]
 Alchemy and alchemists / C.J.S. Thompson.—Dover ed.
   p. cm.
 Originally published: The lure and romance of alchemy. London : George G. Harrap & Co., 1932.
 Includes index.
 ISBN 0-486-42110-4 (pbk.)
  1. Alchemy—History. I. Title.

QD13 .T5 2000
540'.1'12—dc21

2001053804

Manufactured in the United States of America
Dover Publications, Inc., 31 East 2nd Street, Mineola, N.Y. 11501

NOTE

The author wishes to acknowledge his indebtedness to the works of M. Pattison Muir, J. E. Mercer, and A. S. Ridgrove.

CONTENTS

CHAPTER		PAGE
I.	THE DAWN OF ALCHEMY	9
II.	THE ROMANCE OF THE SEVEN METALS	16
III.	THE DIVINE ART IN THE NEAR EAST	25
IV.	THE MYSTERY OF THE EMERALD TABLET	31
V.	THE INFLUENCE OF ASTROLOGY ON ALCHEMY	36
VI.	THE GREEK ALCHEMISTS	41
VII.	ALCHEMY IN CHINA AND INDIA	49
VIII.	THE ARABIAN ALCHEMISTS	59
IX.	THE PHILOSOPHER'S STONE AND THE ELIXIR OF LIFE	68
X.	FAMOUS ALCHEMISTS OF MEDIEVAL TIMES	77
XI.	ENGLISH ALCHEMISTS OF THE MIDDLE AGES	91
XII.	ALCHEMY IN THE TIME OF JOHN GOWER AND GEOFFREY CHAUCER	102
XIII.	THE ALCHEMIST'S LABORATORY AND HIS APPARATUS	108
XIV.	ALCHEMICAL SYMBOLS AND SECRET ALPHABETS	120
XV.	SOME NOTABLE ALCHEMICAL MANUSCRIPTS	134
XVI.	ALCHEMISTS AND ROYALTY	140
XVII.	ALCHEMISTS IN FORTUNE AND MISFORTUNE	150
XVIII.	FAMOUS ALCHEMISTS OF THE SIXTEENTH CENTURY	159
XIX.	CÆTANO AND BORRI	174
XX.	ALCHEMISTS OF THE SEVENTEENTH CENTURY	188
XXI.	BEN JONSON'S "ALCHEMIST" AND OTHER QUACKS	197
XXII.	THE ROSICRUCIANS AND ALCHEMY	203
XXIII.	MORE ALCHEMISTS OF THE SEVENTEENTH CENTURY	209

ALCHEMY AND ALCHEMISTS

CHAPTER		PAGE
XXIV.	THE LAST OF THE ALCHEMISTS	216
XXV.	THE DAWN OF SCIENTIFIC CHEMISTRY	222
XXVI.	ALCHEMY AND GOLD-MAKERS IN MODERN TIMES	233
XXVII.	WHAT WE OWE TO THE ALCHEMISTS	240
	INDEX	245

ILLUSTRATIONS

	PAGE
THE ALCHEMIST	*Frontispiece*
AN ALCHEMIST	18
THE ALCHEMIST	19
HERMES TRISMEGISTUS	26
ANCIENT EGYPTIAN GOLDSMITHS AT WORK	27
AN EGYPTIAN SMELTING AND FUSING METAL	27
AN ALCHEMIST AND AN ASTROLOGER IN CONSULTATION OVER A PROCESS TO BE CARRIED OUT WHEN THE MOON IS IN THE SIGN OF SCORPIO	38
TRIBICUS AND STILL	44
STILLS	45
APPARATUS FOR DIGESTION	46
SYMBOLIC FIGURES	65
STILLS	65
A GRADUATED RECIPIENT	66
SYMBOLIC REPRESENTATION OF THE BIRTH OF THE PHILOSOPHER'S STONE	70
TITLE-PAGE FROM "BASILICA CHYMICA," BY CROLLIUS (1612), DEPICTING THE "GREAT MASTERS AND SYMBOLS OF ALCHEMY"	71
APPARATUS FOR DISTILLING AQUA VITÆ	75
THE STONING OF RAYMOND LULLY	81
AN ALCHEMIST	86
SYMBOLIC FIGURE	87
ALCHEMISTS AT WORK IN A LABORATORY	90
MINIATURE OF THE FIFTEENTH CENTURY DEPICTING AN ALCHEMIST IN HIS LABORATORY	96

ALCHEMY AND ALCHEMISTS

	PAGE
Miniature of the Fifteenth Century representing an Alchemist's Laboratory in which Two Assistants are engaged in various Operations	97
An Alchemist blowing the Fire of a Furnace	100
Plan and Elevation of a Laboratory	109
Tribicus, Still, and other Apparatus described by Synesius in the Fourth Century	110
Stills	110
An Alchemist in his Laboratory	110
The Alchemist	111
Still-heads and Alembics	111
Retorts and Matrasses	112
Pelicans, Cucurbits, and Twins	112
Apparatus for Digestion or Sublimation	113
Athanors	113
Apparatus for Digestion	114
Serpent Condenser	114
Furnaces of various Types	115
Apparatus for Distillation	116
A Retort and Recipient Still for making Spirit of Wine	117
Distilling Apparatus	118
A Hot Still	118
Apparatus employed in a Laboratory in the Late Seventeenth Century	119
Early Greek Alchemical Symbols	122
Still for making Oil of Vitriol	122
Multiple Still for making the Elixir di Mara Vigleosa Virtu	123
Symbolic Representation of one of the "Twelve Keys of the Great Stone of the Philosophers"	126
A Monastic Laboratory	126
Symbolic Figure representing the Green Lion devouring the Sun	127

ILLUSTRATIONS

	PAGE
Symbolic Figure representing Coagulation of the First Stone and its Sublimation	127
Symbolic Figure representing a Process in Alchemy	127
Symbolic Alchemical Figure representing a Process	128
Drawings of Apparatus for Separation and Rectification	128
Symbolic Representations of Operations	129
Secret Alphabets used by Alchemists in the Fifteenth and Sixteenth Centuries	131
Some Alchemical Symbols used during the Medieval Period	132
An Alchemist	136
Gold and Silver represented by a King and Queen	137
Licence granted to John Artec to practise Alchemy and continue his Work of Transmutation	141
Rudolph II, Emperor of Germany, in the Laboratory of his Alchemist	152
The Alchemist	153
Henry Cornelius Agrippa	163
Paracelsus (1493–1541)	166
Ivory Mortar and Pestle carved in Relief with a Representation of a Laboratory	167
Apparatus for Transmutation described by Paracelsus	168
Reputed Autograph of Paracelsus	172
Symbolic Figure representing an Alchemist and his Wife engaged in the Preparation of the Philosopher's Stone	192
An Alchemist and his Wife engaged in various Operations	193
A Mystic Alchemist in his Laboratory	210
Page from a Manuscript on Alchemy, dated 1576, by Christopher von Hirschenberg	211
Alchemists at Work in a Laboratory	211
The Origin of the Pelican	215

ALCHEMY AND ALCHEMISTS

	PAGE
THE MATRASS LIKENED TO AN OSTRICH	215
A LABORATORY IN 1747	224
A LABORATORY	225
DISTILLING PHOSPHORUS IN GODFREY'S LABORATORY	228
A LABORATORY, SHOWING FURNACES, RETORTS, AND VARIOUS APPARATUS USED FOR DISTILLATION IN 1751	230
A LABORATORY, SHOWING FURNACES AND RETORTS USED FOR DISTILLATION IN 1751	231
A LABORATORY	231

ALCHEMY AND ALCHEMISTS

CHAPTER I

THE DAWN OF ALCHEMY

THAT alchemy has appealed to the imagination of man for centuries is evident from the prominent part it plays in the legends and romances of the past.

To the artist the alchemist at work in his dim, mysterious laboratory, with its glowing furnaces and fantastic apparatus, formed an attractive subject for his brush, while the poet found in his romantic and picturesque life a fascinating theme for his pen. A great deal of this attraction was doubtless due to the mystery with which the art of alchemy has ever been surrounded.

Among the early civilizations, from the beginning, practically all knowledge was influenced by mysticism. There was a general belief that everything came from the gods. Everything, from the stars that spangled the dark heavens in their splendour to the terrestrial phenomena with all their natural forces, was the work of the Great Unknown. The goodwill of the mysterious powers of the unseen had always to be propitiated. Thus we find that in alchemy, as in the occult sciences, there was a belief in the teachings of the early mystics which was combined with the primitive idea of invoking the aid of the unseen world by magical and other similar practices.

At its dawn alchemy was regarded as a divine and sacred art, enveloped in mystery, that was only to be approached with reverence. Its adepts held its secrets inviolate, enshrouded their operations with symbolism, and gave their materials

ALCHEMY AND ALCHEMISTS

fantastic names so as to conceal their identity from those outside the mystic cult.

The word 'alchemy' is said to be derived from the Arabic definite article *al* prefixed to the late Greek word *kimia*, a word found in the decree of Diocletian against the old writings of the Egyptians on the *kimia*, or transmutation of gold and silver. Suidas, writing in the eleventh century, says that the word means "the knowledge of Egyptian art, Chemi or Cham or the Black Land, which was the ancient name for Egypt."

Some writers suggest that the word *chymike*, which was first used by Alexander of Aphrodisias to denote work done in a laboratory and was later applied to those who practised alchemy, was the origin of the name given to the art. Such workers chiefly occupied themselves with the preparation of plants and herbs for medicinal purposes, and it has been conjectured that these pursuits developed into the study of metals. On the other hand, the famous Dutch physician Boerhaave, who made a careful study of the etymology of the word, believed that it meant something secret or occult, and that it owed its derivation to the Hebrew *chamaman*, or *hamaman*, 'a mystery,' as it was considered "not fit to be revealed to the populace, but treasured up as a religious secret."

In English literature the earlier methods of spelling 'alchemy' were 'alcamye,' according to a manuscript dated 1377, and 'alcamistre,' used by Chaucer in 1386, while other variants employed in the fifteenth century were 'alcamystere' and 'alcamystrie.'

Baker, who translated Gesner's *Jewell of Health*, is one of the first to spell alchemist with 'y,' in the allusion to a "certain Alchymister in Padua," after which the 'y' and 'e' were both used throughout the seventeenth and eighteenth centuries.

There is a curious tradition regarding the origin of the word recorded in *Hermes and the Holy Writings*, where we read that "a race of giants was the result of a union between certain

THE DAWN OF ALCHEMY

women and some spirits, who because of this were cast out from heaven and condemned to eternal exile. The books in which they taught their arts were called *chema*, and later this name became applied to the art itself."

Tertullian and other early writers mention this legend, the former observing that "these abominable and useless books were the teachings of the fallen angels to their wives on the art of poisons, the secrets of metals, and magical incantations."

Clement of Alexandria, in alluding to the tradition, says of this mysterious race: "They laid bare the secrets of metals, they understood the virtues of plants and the force of magical incantations, and their learning even extended to the science of the stars." Thus they combined a knowledge of alchemy, magic, and astrology.

Though alchemy may have begun with the study of metals, it soon became associated with magic, pharmacy, and the influence of the stars.

We know from later writers that the Arabian physicians attached the greatest importance to the influence of alchemy on the discovery of new drugs.

The fundamental theory underlying the art of alchemy is more obscure than its history, and its elucidation is rendered more difficult owing to the symbolism with which all its operations and practices were surrounded. The supposition that it had its origin in the experiments carried on by the ancient workers in metals is improbable, as there is evidence of its practice at a much earlier period. It is possible that the quest for the secret by means of which human life might be prolonged, afterward called the 'Elixir,' was the first phase of its origin, as among both the Chinese and the Hindus, whose knowledge of alchemy goes back to a period of great antiquity, the transmutation of the baser metals into gold was but a secondary aim.

Among Western nations alchemy developed into a philosophical as well as an experimental science; therefore in endeavouring to find the principles underlying its origin we must consider it from both the mystical and the physical points of view.

It is contended by some that alchemy originated in the attempt to demonstrate the applicability of the principles of mysticism to the things of the physical realm. This idea brings into harmony the physical and transcendental theories of alchemy and the various conflicting facts advanced in favour of each. It also affords some explanation of the existence later of alchemists of both types and the general religious tone of many of the alchemical writings.

Certain mystics considered that alchemy was concerned with the soul of man and that its chief principles were allegories dealing with spiritual truths, and in support of this idea pointed out that in the early ages the Divine Art was practised only by the priesthood. Certainly we have no mention of the transmutation theory among Western nations until about the seventh century of the Christian era.

From the dawn of science we find a belief that the four great primal elements produced, by reacting one upon the other, certain principles which in turn were the causes of the metals. Thus the elements were supposed to produce what was called their seed, and the principles matured and perfected them. Every class had its own seed, and to obtain this was to possess the power of producing the thing that sprang from it.

Aristotle's theory of the four elements—water, air, fire, and earth—held sway for centuries, and influenced the teaching of science down to the Middle Ages, although the early alchemists did not regard the metals as elements. Let us consider how this elemental theory arose.

From the sea came the gods who instructed mankind in the sciences and learning; this tradition is embodied in the earliest Babylonian records. The Hindus had similar legends, which connected some of their deities with the sea; and from these it is probable that water had a place in man's earliest system of philosophy. Water was the solvent of all things; it entered into

THE DAWN OF ALCHEMY

the composition of all animal bodies, and was necessary to life itself.

Rain was considered to be caused by the condensation of clouds, which in their turn were formed by the condensation of air; and there was a general belief, which persisted for centuries, that water might be converted by fire into air. There was also a theory that water might be transmuted into air.

After water, air, on which Anaximenes (sixth century B.C.) was the first to comment, came next in importance. It was regarded as the universal bond of nature which held in itself the substantial principles of all natural things. "The elementary air was believed to be the universal world spirit," says an early writer, "for the oracles assert that the impression of characters and other divine visions appear in the ether."

The idea of fire as a primal element is thought to have originated in the fire- or sun-worship of the early civilizations, and it was added to the elements by Heraklitus of Ephesus. It was regarded as the cause of all motion, and consequently of all change in nature. Its universal centre was believed to be in the heavens and its locality in the earth. It was the principle of all generation and the primal source of all forms in itself, boundless and inscrutable. It was considered to be the highest active and elastic element, one in essence, though manifested in three forms—celestial, subterranean, and for culinary and other terrestrial uses.

Pherekides considered earth also to be a primal element, but it was regarded as passive, although it was believed to be the centre in which all the others operated. It was the final receiver of all the influences of the heavenly bodies, the common mother from which all things sprang, whose fruitfulness was produced by the threefold operation of fire, air, and water.

According to Zoroaster, the Creator made the whole world of fire, water, and earth, and all-nourishing ether. Thus water, earth, air, and fire were recognized as the four primary elements by Aristotle and Plato. At a later period it was conceived that each element possessed two qualities, and from this fire came to be regarded as more hot than dry, air as more wet than hot, water as more cold than wet, and earth as more dry than cold.

These ideas eventually came to be applied to the seven metals —gold, silver, mercury, iron, lead, copper, and tin—which at first were regarded more as properties than substances. Thus gold, the most precious and rarest of all metals, was the most perfect substance of its colour, a property or quality supposed to be latent in all metals. There was also a belief that all metals but gold were Nature's failures, and that by adding something to the baser metals the golden property hidden in them might be found and the design of Nature completed.

The idea that matter is composed of small particles or atoms that are in a state of ceaseless motion is to be met with in Hindu and Phœnician philosophy, and was taught also by the Greeks. Anaxagoras considered every atom to be a world in miniature, and that the living body was composed of atoms derived from the elements which sustain it. According to other theories, atoms were variable not only in size, but in weight; they were impenetrable, their collision giving them an oscillatory movement which they communicated to adjacent atoms, and this movement they in their turn passed on to others more distant.

Man has not yet been able to divide an atom, and the theory of the philosophers of ancient times as to the oscillatory movement of the atom was the genesis of our knowledge of the ether to-day.

As the value of metals became understood and their use in the arts realized gold came to be regarded as the rarest and most precious of them all, and those who worked in them began to seek other methods of obtaining it. They asked themselves how the necessary change in other metals could be brought about. Surely, they thought, some medium could be found whereby the latent property believed to exist in these metals could be brought out, thus changing them into gold. The changes that were wrought with metals by the action of fire, air, and water soon became obvious to those who worked in them, and the value of mixing one with another in alloy also became apparent.

From these things men gleaned their first knowledge of chemical phenomena, and thus probably the idea of transmutation came to form part of the earliest systems of philosophy.

CHAPTER II

THE ROMANCE OF THE SEVEN METALS

THROUGHOUT the writings of the early alchemical philosophers there is constant reference to the mystic connexion between the seven metals—gold, silver, copper, mercury, lead, tin, and iron—and the seven planetary bodies—the Sun, the Moon, Venus, Jupiter, Mercury, Saturn, and Mars.

The origin of this association takes us back to a very early period when the astrologers of Babylonia formed a cult of their own that became known as the Chaldean art. This spread to Egypt, and thence was transmitted to the Greeks and so on to the Romans, who associated many of their deities with the planets. The belief that the seeds of the metals were in the earth and that their formation or growth was fostered by the influence of the planetary bodies is therefore of great antiquity. Proclus in his commentary on *Timæus* says:

> Natural gold and silver, as well as all the other metals and like other substances, are produced in the earth under the influence of the celestial divinities and their effluvia. The sun produces gold, the moon silver, Saturn lead, and Mars iron.

It was believed that no planet could undergo a modification without arousing a corresponding sympathy in the metal. This sympathy or bond was supposed to be transmitted by invisible minute bodies which proceeded to the metals from the planets, each of which had a day in the week on which it manifested its influence over its particular metal. Thus, gold was dedicated to Sunday, silver to Monday, iron to Tuesday, and so on.

So strong was the belief in the growth of metals that down to the sixteenth century mines were sometimes closed for a while in order that the supply of metal might be renewed.

THE ROMANCE OF THE SEVEN METALS

According to a work of Basil Valentine, said to have been written at about that period,

> All herbs, trees, and roots, and all metals and minerals, receive their growth and nutriment from the spirit of the earth, which is the spirit of life. This spirit is itself fed by the stars, and is thereby rendered capable of imparting nutriment to all things that grow and of nursing them as a mother does her child while it is yet in her womb. The minerals are hidden in the womb of the earth and nourished by her with the spirit which she receives from above.

From this supposed close association between the planets and the metals probably came the spiritual connexion that existed between alchemy and astrology in the early ages. To each of the seven metals was assigned the symbol of the planet which was believed to influence it. Thus, the sign used for the sun became the alchemist's symbol for gold, while that used for the moon became the symbol for silver, and similarly with the other planets.

Without pursuing these old doctrines further, let us consider some of the traditions connected with the history of the seven metals.

Gold, the most precious of all, has ever had an attraction for man. It is probable that it was one of the earliest metals of which he made use, and may, indeed, have been the first he discovered, as it was to be found free in nature, in rocks as well as in the sands of rivers. In Ethiopia and Nubia it must have been known at a period of great antiquity, and there is evidence that quartz-crushing and gold-washing were known in Egypt before 2500 B.C.

From the Middle Ages it was believed to possess great medicinal virtues. As Chaucer observes of his Doctour of Phisik in the Prologue to *The Canterbury Tales*,

> Gold in phisik is a cordial,
> Therefore he loved gold in special.

In the sixteenth century Paracelsus recommended gold to purify the blood, and states that if put into the mouth of a newly born babe it will prevent the devil from acquiring a power over the child. In the seventeenth century it became included as an official remedy in many of the Pharmacopœias of Europe, and was considered to be superior to mercury in its action in the treatment of many diseases.

Crocus solis, the preparation of gold generally used, was made by dissolving gold in nitro-hydrochloric acid, and after distilling the solution with water and precipitating it with solution of potass (potash) the resulting powder was washed and dried. The stannate of gold (purple of Cassius) was prescribed internally, and when made into a solution was believed by some to be the Elixir of Life.

The metal itself was employed in the shape of an amalgam made with mercury and gold-leaf, which was administered in the form of a syrup or tincture. The bromide of gold was used in medicine down to recent times. In alchemical symbolism and allegory gold was represented as a king.

Silver, which was originally called white gold, entered into the composition of the electrum of the Greeks, and, alloyed with gold, formed the metal from which some of the earliest-known coins were struck. Silver nitrate, probably the first salt prepared from the metal, is said to have been discovered by Jábir-ibn-Háyyan (Geber) about the eighth century. It was known in the Middle Ages as *lapis infernalis* and afterward as lunar caustic, and was first prepared in sticks, now commonly called caustic, by Glaser, an apothecary in Paris in the seventeenth century. Silver was associated with the moon, and for this reason was regarded as a potent remedy for all diseases affecting the brain. It was used by the Arab physicians in the treatment of vertigo and falling-sickness. The employment of pills of silver continued down to the eighteenth century, a favourite form of administration being *pilulæ lunares*, or Pills of the Moon, which were composed of silver nitrate combined with opium, musk, and camphor. Another preparation, known as tincture of the moon, consisted of a solution of silver nitrate

AN ALCHEMIST
From an engraving after Stradano

THE ALCHEMIST
From an engraving after Brueghel

THE ROMANCE OF THE SEVEN METALS

mixed with a little copper to give it a bluish tint. The metal was represented by the alchemists as a queen, and its symbol was a crescent moon.

Copper, which was associated with the planet Venus, was known and used before iron, and was smelted from the ores by primitive man. Its ductility rendered it particularly adaptable for the making of utensils and implements, and later it formed a component part of the bronze employed for weapons and knives. As the copper ores were frequently found associated with other metals it was probably soon discovered that by alloying it could be rendered hard enough to fashion into tools. It was known to the Greeks in the time of Homer, for he tells us that the shield of Achilles was composed of gold, silver, tin, and copper, while the arms of the heroes were of copper. Bronze was used by the Egyptians as early as 2000 B.C., and was employed for making vases, statuettes, mirrors, and arms, and the alloy used consisted of from 80 to 85 per cent. of copper, with between 20 and 15 per cent. of tin.

The metal-workers in ancient times obtained the red and black copper oxides by heating copper to redness and allowing it to cool in the air. They distinguished between the scales which fell off during the cooling and those that could be obtained by heating the metal, and these oxides were used for colouring glass. The oxy-acetate of copper, or verdigris, was known and used as a pigment at least five thousand years ago, while later it was employed by the Egyptians, Greeks, and Romans as a remedy for various affections of the eyes. The Egyptians prepared it by covering plates of copper with the refuse of grapes after the juice had been taken.

At a very early period mercury was known in China and India, where it was found native or obtained by heating cinnabar, the sulphide, with iron filings in an earthen vessel to the top of which a cover was sealed with clay. The iron decomposed the sulphide, and the liberated mercury was volatilized and condensed on the cover, whence it was collected. This early crude form of distillation became one of the chief operations among the alchemists for separating the volatile from fixed substances.

Mercury is mentioned both by Aristotle and Theophrastus about 325 B.C., and was originally called 'quicksilver' when found in the liquid state. About A.D. 50 it is mentioned by Dioscorides, who calls it 'hydrargyrum,' or 'liquid silver,' and states that it was frequently confused with minium, or red oxide of lead. It was originally thought that cinnabar, the sulphide, and native quicksilver were two distinct substances. The important use of mercury for extracting gold from its matrix and other metals goes back to an early period, and Pliny records that quicksilver was employed in his time for separating the noble metals from earthy matter. The crushed gold-quartz was shaken up with mercury, which dissolved out the gold; the amalgam of gold and mercury thus obtained was then squeezed through leather, which separated most of the mercury. The solid amalgam was then heated to expel the mercury that was left, and the pure gold remained. It is probably owing to this power of mercury over gold that the alchemists came to regard it as a very important metal.

The Arabian alchemists were the first to investigate the properties of mercury, and it is to Jábir-ibn-Háyyan that we owe the earliest description of the red oxide, and also of the perchloride of mercury commonly known as corrosive sublimate. Until about the eighth century mercury was regarded as a powerful poison. It is so described by Galen, but Avicenna, the Arabian physician, observed that quicksilver in the metallic form could be swallowed without ill-effect and that it passed through the body unchanged. Following the old theory of the alchemists, Jábir-ibn-Háyyan laid it down that all metals consisted of sulphur and mercury in different proportions and of different degrees of purity. Sulphur, he believed, caused the alterations in metals when heated, while mercury imparted lustre and malleability as well as other metallic properties.

THE ROMANCE OF THE SEVEN METALS

The red oxide of mercury subsequently played an important part in the history of chemistry. The French chemist Charas, who lived in the seventeenth century, called it *arcana corallina* on account of its colour. Robert Boyle prepared it by boiling mercury in a bottle which was fitted with a stopper provided with a narrow tube by which air was admitted. The product was afterward called 'Boyle's Hell,' because it was believed that it caused the metal to suffer extreme agonies during the process. It was from the red oxide that Priestley first produced what he called 'dephlogisticated air,' which we now know as oxygen —a discovery which opened a new era in the history of chemistry.

The internal administration of the salts of mercury was popularized in the sixteenth century by Paracelsus, who advocated the use of the perchloride, the oxide, and the nitrate of the metal. It was employed as a remedy in the form of fumigations, frictions, ointments, and plasters. John de Vigo, of Naples, physician to Pope Julius II, devised a mercurial plaster. Quicksilver girdles, or belts prepared with mercury and the whites of eggs, also became popular as a remedy for itch.

It was recognized in the sixteenth century that mercury in the state of minute subdivision had a distinct physiological effect, and attempts were made to extinguish or 'kill' the metal by trituration and so render it efficient for internal administration. A preparation of this kind known as 'grey powder,' made by triturating mercury with chalk, is still largely used in medicine.

Braun, of Petrograd, was the first to solidify mercury, which he effected by placing a thermometer in a mixture of snow and nitric acid. The mercury sank with great rapidity, and when Braun examined the bulb he found it contained a metallic mass which could be hammered like lead.

Another important event in the history of mercury was the experiment made in 1643 by Torricelli, who determined the pressure of the atmosphere as equal to 30 inches in a column of mercury. This discovery resulted in the invention of the mercurial barometer. Fahrenheit introduced the use of mercury in thermometers about 1720, and Priestley was the first to use the metal as a sealing agent when working with gases that were soluble in water.

In the arts mercury was used by the Venetians in the preparation of tin amalgam for silvering mirrors as early as the sixteenth century.

Iron, which was associated with the planet Mars, is said to have been known from at least 1537 B.C. It was tempered by heating to redness and then plunging the metal into cold water. There are representations of a bellows being used in smelting, together with a furnace and crucible, in an Egyptian wall-carving dating from 1500 B.C. Iron rust was recognized as a tonic and styptic in classical times. Homer refers to the rust of the spear of Telephus being used to heal wounds which the weapon had inflicted. The traditional Melampus (1380 B.C.), who is said to have studied Nature while tending his sheep on the mountain-sides, and afterward acquired the reputation of being a great healer, when called to treat Iphiclus, King of Phylacea, who was suffering from extreme weakness, instructed him to take the rust of iron in wine. The remedy proved eminently successful, and iron wine is used for its tonic properties to this day.

Iron has ever been associated with strength, and its supposed connexion with Mars probably influenced its early use in medicine, especially as its value in chlorotic diseases was doubtless soon observed. Dioscorides (A.D. 40) refers to its astringent properties, and recommends its use in certain cases of hæmorrhage, while Celsus and Pliny both allude to the value of iron in water or wine as a remedy for dysentery and enlargement of the spleen.

The salts of Mars, as the preparations of iron were called, became popular in the seventeenth and eighteenth centuries mainly through Sydenham and Willis, two famous English physicians, who advocated the use of iron in the treatment of debilitated conditions. Willis originated and prescribed a secret

THE ROMANCE OF THE SEVEN METALS

preparation of iron which was known as 'Dr Willis's Preparation of Steel' and, according to a contemporary, "was the best preparation of any that iron can yield us."

Crocus martis, the sesquioxide, *æthiops martial*, the black oxide, and *flores martias*, the ammoniated chloride of iron, the old names for which suggest their connexion with the god of war, were all popular remedies in the seventeenth century.

Early in the eighteenth century a secret preparation of iron known as the 'Golden Drops of General La Mothe' became extremely popular in France, and a similar remedy called 'Bestucheff's Nerve Tincture' had a great vogue in Germany and Russia. They were believed to be solutions of gold, and were recommended for their marvellous restorative properties. Although the price of a small bottle containing about four teaspoonfuls was a livre, they became so famous that Louis XV sent two hundred bottles to the Pope as a particularly precious gift. Later Louis granted La Mothe a yearly pension of 4000 livres for the right of making the preparation for the Hôtel des Invalides. When the formula was subsequently published the much-renowned 'Golden Drops' were found to consist of tincture of perchloride of iron and spirit of ether.

Lead was known to the Egyptians at a very early period, and they employed both the protoxide, or litharge, and minium, or red lead, as pigments. White lead, or *cerussa*, the carbonate, they obtained by exposing sheets of the metal to the fumes of vinegar in a warm place. It was associated with the planet Saturn, and many of the preparations of the metal took their name from the planet. Thus, the 'Magistery of Saturn,' which was prepared by precipitation from a solution of the acetate of lead with potassium carbonate, became famous in the seventeenth century. It formed the chief ingredient in the renowned 'Powder of Saturn,' which was originated by Mynsicht and largely used in the treatment of asthma and phthisis. Later, about the middle of the eighteenth century, a revival in the use of lead in medicine was brought about by Goulard, a surgeon of Montpellier, who wrote a treatise on the 'Extract of Saturn' which made preparations of lead celebrated throughout the world, and its employment as an external application continues at the present day.

Tin, which was associated with the planet Jupiter, was known to the Egyptians as far back as 2000 B.C. The Phœnicians, who were the earliest-known traders in the metal, obtained it first from India and Spain and afterward from Britain, where the tin-mines of Cornwall have been worked from a remote period of antiquity. It was originally regarded as of higher value than copper. The alchemists first prepared the salts of the metal by calcination, upon which vinegar was poured, and they found that by heating them together they obtained a crystalline salt. They made the oxide by strong ignition with charcoal.

The chief salts of tin were known as *sal jovis*, the nitrate or chloride, and *calx jovis*, the binoxide, the former of which was used medicinally as a vermifuge. *Aurum musivum*, well known to the alchemists, was prepared by combining tin and mercury into an amalgam and then distilling it with sulphur and sal ammoniac. The product, a beautiful golden metal of crystalline structure and brilliant lustre, was highly esteemed and recommended in the treatment of fevers and hysteria; it was also said to possess sudorific properties.

Much more might be written concerning the seven metals on which the early alchemists laboured for centuries, but this brief outline will serve to show how far they succeeded in preparing certain salts that have proved of great value to mankind, many of which are in use at the present time.

CHAPTER III
THE DIVINE ART IN THE NEAR EAST

THE country in which the Divine Art first had its birth cannot be stated with exactitude, and the genesis of alchemy has been variously ascribed to Egypt, Babylonia, India, and China. That there was some knowledge of the science in both the Near and Far East at an early period there is little doubt, but no documentary evidence is available until about 2500 years before the Christian era, from which time we have record of the working of metals in Egypt.

Although we are told that the Egyptian priests were initiated into the mysteries of the Divine and Sacred Art in the temples, and were afterward able to imitate the work of the deities, little is known of the nature of their operations. The art was kept secret and almost exclusive to the priesthood, and beyond kings and their sons no one was allowed to be instructed in its mysteries. That it had a close alliance with magic and astrology is shown in some of the stories recorded in Egyptian papyri of a later period.

There is an ancient tradition that Hermes Trismegistus, called "the Thrice Great," was the originator of alchemy in Egypt. This legendary personage is supposed to have been an Egyptian priest who flourished about 2500 years before Christ. He may have been connected with the god Thoth or perhaps was an emanation from that deity, who on account of his great intellectual powers became one of the three great triune gods of Memphis. Hermes is referred to on the Rosetta Stone as the "Great and Great," and his name is perpetuated in connexion with chemistry to the present day in the term applied to the method of enclosing a body in a glass tube by fusion or sealing, which in ancient times was called "Hermes, his seal," or, as we now express it, hermetically sealed.

From early sculpture and wall-carvings that have been discovered in Egypt there is evidence of a considerable knowledge of the methods of working metals and grounds for believing that in all probability the Egyptians were also the first people to apply that knowledge to everyday needs. They had gold from the land of Ophir, Nubia, and the region of Meröe. The gold-mines of Ethiopia and Nubia were extensively worked, and Diodorus Siculus alludes to the number of slaves employed for this purpose. The finely ground gold-ore was washed out, and the heavy residue melted. In the time of Rameses II the mines are said to have yielded gold to the value of £125,000,000 sterling per annum.

HERMES TRISMEGISTUS
From an ancient Egyptian stone carving

The metal was also found in a matrix of quartz, and obtained by crushing and washing. Representations of these processes, showing gold-washing, fusion, and the weighing of the metal, are incised on a tomb dating about 2500 B.C. at Beni Hassan. The use of a furnace and the blow-pipe is also depicted, the raised portion of the former being used for the purpose of concentrating the heat upon the crucible on the principle of the reverberatory furnace.

There is further evidence given in the Bible of the richness of the country in the precious metal, for it is recorded [1] that the Queen of Sheba brought much gold and precious stones and

[1] 1 Kings x, 10, 14.

IN THE NEAR EAST

gave to King Solomon 120 talents, a sum equivalent to £240,000. The navy of Hiram also brought gold from Ophir, and the

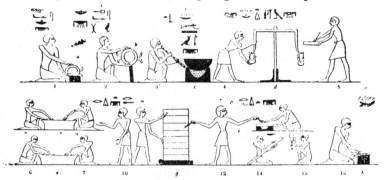

ANCIENT EGYPTIAN GOLDSMITHS AT WORK
The processes represented are making jewellery, blowing the furnace for melting gold, weighing the gold, washing gold, and preparing the metal before working it.
From an incised carving found at Beni Hassan
Wilkinson, "*Ancient Egyptians*"

weight of gold that came to Solomon in one year was 666 talents, estimated at about 37,296 pounds avoirdupois.

The use of gold for making jewellery dates back to a remote age, and the perfection to which the art of the goldsmith attained may be judged from the beautiful specimens of the craft discovered in many Egyptian tombs. The Egyptians produced most elaborate and exquisite ornaments of beaten gold in the shape of flowers, leaves, pectorals, bracelets, rings, and statuettes. They drew gold wire for the purpose of embroidery, and also used the metal for inlaying. They practised the art of enamelling, a beautiful specimen of which, in the form of an amulet, was found on the mummy of

He is using a blow-pipe and small furnace with cheeks to confine and reflect the heat.
From an incised carving found at Thebes
Wilkinson, "*Ancient Egyptians*"

Queen Aahotep, who lived about 1700 B.C. They acquired great skill in colouring glass with various metallic oxides, and made beads and imitation precious stones as early as 1475 B.C. In one of the papyri are several interesting formulæ for making artificial emeralds and hyacinths; these are said to have been taken from the *Book of the Sanctuary*. In another papyrus are three recipes for making silver.

The art of dyeing was known and practised in ancient Egypt, and the Egyptians understood the effect of acids on some colours, and were acquainted with mordants for fixing the colouring matter in the fabrics. They made and dyed linen over five thousand years ago, using indigo which they obtained from India. They employed sulphur, alum, antimony, copper, cobalt, verdigris, zinc oxide, white lead, and arsenic in the arts, the latter in the form of realgar, the red disulphide, and orpiment, the yellow trisulphide of the metal.

Of their knowledge of alchemical operations very little is known. Zosimus mentions that, apart from the knowledge contained in the hermetic books, all the learning of the priests was secrets which it was forbidden to reveal and was enveloped in mystery.

In his treatises John the High Priest states that Thebäis, Heracleopolis, Lycopolis, and Elephantine, all great cities of Egypt, were strongholds of the priesthood, where probably they purified gold and other minerals and also pursued endeavours to make them. Zosimus observes that there were strong reasons for treasuring alchemy as a religious secret, and even after the removal of the seat of the Roman Empire to Constantinople the art appears to have been regarded as a sacred mystery to be revealed only to the priesthood and jealously guarded by them.

From the frequent communication between Babylonia and Egypt in ancient times it is natural that we should find evidence of the practice of the hermetic art among the early races that inhabited Assyria and Mesopotamia. A translation of a series of texts from clay tablets of the period of Assur-bani-pal, who flourished between 668 and 626 B.C., shows that the Assyrians

IN THE NEAR EAST

at that early period practised the art of smelting metals and knew how to make glass and colour it by means of metallic oxides.

Recipes are recorded for making blue-glaze with sand, alkali ash, and gum styrax. Arsenic, red alum, cinnabar, and saltpetre were employed in the manufacture of purple glass, and rust, saltpetre, and oxide of tin were used for green crystal.

Another text reads:

> When thou settest out the plan of a furnace for minerals thou shalt choose out a favourable day in a fortunate month, and thou shalt set out the ground-plan of the furnace. While they are making the furnace thou shalt watch them and shalt work thyself. Thou shalt bring in embryos [born before their time]; another, a stranger, shall not enter, nor shall one that is unclean tread before them; thou shalt offer the due libation before them; the day when thou puttest down the mineral into the furnace thou shalt make a sacrifice before the embryos; thou shalt set a censer of pine incense.

This allusion to the choice of a favourable time for carrying on an operation is especially interesting as showing the connexion of astrology with the process of smelting. Notable also is the propitiatory libation and incense offered to the gods. The use of the *fœtus*, or embryo, probably originated in the Assyrian belief that the spirit of an incomplete being must be propitiated, as it might have some mysterious or malign influence over the substances about to be used in the operation.

Olympiodorus alludes to a similar tradition common among the Greek alchemists, who believed that "certain demons were jealous with regard to the making of some recipes."

Another text deals with the making of oxide of copper, which the Assyrians used for colouring glass blue. It is directed to be prepared as follows :

> Thou shalt put 10 mana of copper into a clean melting-pot; thou shalt put it down into the furnace which has been let grow hot . . .; thou shalt keep a fierce fire burning until the glass-copper fuses. Thou shalt beat . . ., thou shalt open . . . until the Zuku . . . glass . . . and thou shalt spread the copper on the roof.

In another text various chemicals used for colouring glass are recorded. Thus, for making red glass gold and oxide of tin (probably the purple of Cassius) were employed, together with antimony and saltpetre; for purple-tinted glass oxide of manganese, and for red-purple glass copper scales were used.

Several recipes are given for making artificial precious stones, including carnelian, spangled redstone, or aventurine, and also for making opaque alabaster. Orpiment, the yellow trisulphide of arsenic, was employed as a paint for the eyes, and a "marrow of arsenic" is mentioned, which appears to have been a preparation for giving the effect of gold, for it was to be employed for redecorating a crown of alabaster. The process of overlaying one metal with another was known at this early period, for in another text allusion is made to the casting of a bronze vessel which was to be overlaid with silver. Among other pigments used in the arts mention is made of vermilion, lapis lazuli (ground to powder), malachite (ground to powder), and white lead.

The art of dyeing and the method of tempering iron were known to both the Babylonians and the Assyrians, and it is probable that from them the Phœnicians, who were the intermediaries between Babylon and Egypt, acquired their knowledge of these arts, the traditions concerning which were passed to the Arabs and the Persians.

CHAPTER IV

THE MYSTERY OF THE EMERALD TABLET

AN atmosphere of romance and mystery surrounds the tradition of an emerald tablet or table that is said to have been discovered in the tomb of the legendary Hermes. It is first mentioned in Western literature in a treatise attributed to Albertus Magnus called *De Mineralibus*, written in the early part of the fourteenth century. In this manuscript it is stated that the tomb of Hermes was discovered by Alexander the Great in a cave near Hebron, and that in the tomb was found a tablet of emerald, taken from the hands of the dead Hermes by Sarah, the wife of Abraham. On this were inscribed in Phœnician characters the precepts of the Great Master concerning the art of making gold. The Hermes alluded to is doubtless intended to mean the traditionary Hermes Trismegistus mentioned in Chapter III.

There are many translations of the inscription supposed to have been found on the tablet, and these in varied Arabic and Latin forms have been carefully studied by Ruska.[1] The earliest forms of the text are in Arabic, and the following is a translation from an Arab collection of commentaries of the early twelfth century known as *The Emerald Table of Hermes*:

> True it is, without falsehood, certain most true. That which is above is like to that which is below, and that which is below is like to that which is above, to accomplish the miracles of one thing. And as in all things whereby contemplation of one, so in all things arose from this one thing by a single act of adoption.
> The father thereof is the Sun, the mother the Moon.

---

[1] Julius Ruska, *Tabula Smaragdini* (Heidelberg, 1926).

The wind carried it in its womb, the earth is the source thereof. It is the father of all works of wonder throughout the world.

The power thereof is perfect.

If it be cast on to earth, it will separate the element of earth from that of fire, the subtle from the gross.

With great sagacity it doth ascend gently from earth to heaven. Again it doth descend to earth and uniteth in itself the force from things superior and things inferior.

Thus thou wilt possess the brightness of the world, and all obscurity will fly far from thee.

This thing is the strong fortitude of all strength, for it overcometh every subtle thing and doth penetrate every solid substance.

Thus was this world created.

Hence will there be marvellous adaptations achieved of which the manner is this.

For this reason I am called Hermes Trismegistus because I hold three parts of the wisdom of the whole world.

That which I had to say about the operation of Sol is completed.

What is the meaning of this enigma? Albertus Magnus, Roger Bacon, and other philosophers of the Middle Ages sought to solve it, but their comments only point to a vague doctrine of correspondence between heaven and earth, so that inanimate nature answers to the planets and the heavenly bodies. It obviously emphasizes the dependence of all earthly things on the sun, thus following the idea of Aristotle that man is generated from man and the sun. It refers to the action of the moon upon the earth, the action of fire on a solid body, causing distillation or sublimation, and the subsequent solution of a rarer liquid. It is, indeed, a brief summary of the principles of change in nature and the foundation of alchemical doctrine, and shows the close connexion between alchemy and astrology.

One of the earliest doctrines of astrology was a belief in a mysterious emanation from the heavenly bodies which influenced man's life in health and disease, and also affected all minerals, plants, and flowers, their properties being derived from the sun, the moon, and the planets.

Legends of the discovery of ancient stone tablets or documents are not infrequent; another is provided by the story of the finding of the famous book on magic known as *The Key of Solomon*, which, according to tradition, was discovered secreted in an ivory casket in a tomb.

In the account of the emerald tablet given by Roger Bacon in the *Secretum Secretorum* it is stated that "These precious sentences of Hermes were found by Galienus Alfachim the physician, on a plaque of emerald in a cave, clasped in the hands of the corpse of that mysterious legendary figure Hermes Trismegistus, The Thrice Great." The reader is exhorted "to preserve the strictest secrecy from all except men of goodwill, this treasured text, even as Hermes himself had hidden it within the cave."

Another instance of a similar discovery is the story respecting the treatise entitled *Concerning the Seven*, attributed to Alexius Africanus, in which the seven herbs connected with the seven planets are named. This document is said to have been found enclosed within a monument with the bones of the first King Kyrannides in the town of Troy.

Several early historians record that the lore of the Egyptians was preserved in the stelæ of their temples. Iamblichus, in the fourth century, mentions "ancient stelæ of Hermes in which all science was written down"; while Olympiodorus, in the sixth century, says, "The secret of the mystic art is inscribed on the obelisks in hieroglyphics."

The tradition that the text was inscribed on an emerald may have arisen from the fact that in Græco-Egyptian times the name was applied to any green stone.

It may be well to quote another and freer translation of this historic text; it can be judged more clearly from this that the writer designed to teach the doctrines of alchemy that were common in the early Christian era.

> I speak not fictitious things, but that which is certain and most true. What is below is like that which is above, and what is above is like that which is below to accomplish the miracles of One

Thing. And as all things were produced by the One Word of One Being, so all things were produced from the One Thing by adaptation. Its father is the Sun, its mother the Moon, the wind carries it in its belly, its nurse is the earth. It is the father of all perfection throughout the world. The power is vigorous if it be changed into earth. Separate the earth from the fire, the subtle from the gross, acting prudently and with judgment. Ascend with the sagacity from the earth to heaven, and then again descend to the earth and unite together the powers of things superior and things inferior. Thus you will obtain the glory of the whole world and obscurity will fly far from you. This has more fortitude than fortitude itself, because it conquers every subtle thing and can penetrate every solid. Thus was the world formed. Hence proceed wonders which are here established. Therefore I am called Hermes Trismegistus, having three parts of the philosophy of the whole world. That which I had to say concerning the operation of the Sun is completed.

The authorship of this remarkable message still remains a mystery, although philosophers have laboured for centuries to prove its authenticity and to interpret its cryptic words. In the Middle Ages it was regarded as a marvellous revelation full of sublime secrets of great importance to mankind, but what these secrets were none was able to reveal.

Ferguson enumerates forty-eight treatises and commentaries on the Emerald Tablet, and remarks that we cannot well ignore it—less perhaps now than ever in view of the discovery of Egyptian writings like the medical *Papyrus Ebers*, which he calls an hermetic treatise of 1550 B.C., a date coinciding with that assigned to Hermes by Lambeck. Other researches have shown that the belief in a person or persons of the name of Hermes has been so widespread and persistent that the whole Hermes legend forms a legitimate subject of inquiry as to its origin.

The text is certainly not modern; it has been assigned to Hermes from the first, and its significance does not lie on the surface. It is a profound mystery and remains a great puzzle. Everything concerning it remains a problem; its legendary and romantic discovery, its author—whether one of the several per-

MYSTERY OF THE EMERALD TABLET

sonages of the name of Hermes or an anonymous writer who ascribed it to him to give it authority—and its possible connexion with so-called hermetic writings of an earlier time. De Sacy was of the opinion that the Emerald Tablet was the work of Apollonius of Tyana, but gives no grounds for his conclusion. The story of its discovery may be a myth, but we must remember that the earliest Egyptian papyri dealing with medicine, which are believed to date from 1550 B.C., were found reposing between the legs of a mummy. The most that can be hoped for is that some future discoveries may lead at least to a plausible theory, if not to perfect certainty, regarding its origin.

CHAPTER V
THE INFLUENCE OF ASTROLOGY ON ALCHEMY

THE study of astrology goes back to the earliest period of the history of man. It formed the basis of many of the ancient mythologies and religions, and among the early races of the world there appears to have been a universal belief that the planetary bodies were the disposers of the affairs of humanity. As primitive man gazed into the starry vaults of the heavens he no doubt gradually realized that certain stars upon which his fate depended accompanied the seasons. He may have reasoned that if they ruled his fate they also governed his body, and it is conceivable that he thus endowed them with divine influence. The first arbitrary division of time, the Zodiac, is generally attributed to the Sumerians, who were among the earliest people to study the stars, and some of the symbols by which they were represented were afterward recorded by the Babylonians on their boundary-stones.

The Egyptians from ancient times had a considerable knowledge of the heavenly bodies, as evidenced in some of their sculptures still extant. The earliest Zodiac known is one found near the Tigris opposite Baghdad, which is believed to date from 1320 B.C. It represents ten out of the twelve signs and ten out of the thirty-six decans, each sign being divided into three decans. The Chinese claim to have studied astrology from the time of the Emperor Hoang Ti, who flourished about 2670 B.C.

How the twelve Zodiacal signs became associated with the various figures and animals used to represent them has long been a matter of conjecture. Primitive man may have noticed

THE INFLUENCE OF ASTROLOGY

some resemblance to animal and other forms in the constellations and named them accordingly, as he did with other phenomena of nature. The Egyptians believed that animals were the guests of the gods and held them in high esteem. They endowed them with thought, reason, and passion, and this also may have contributed to their association with the Zodiac.

Certain celestial influences were believed to emanate from the thirty-six decans of the signs, and the mysterious effect that they exercised on the human body was thought to be due to a subtle ether shed by the heavenly bodies on the earth, that affected not only mankind, but also animals, plants, and mineral bodies. Certain parts of the human body were apportioned to each sign, and the Zodiacal original of the part affected was invoked. This may have arisen from the custom of the Egyptians of placing different parts of the body under the protection of special divinities, both in sickness and in health.

The early astrologers believed that each organ of the human body was formed by the action of certain media that existed in the universe; thus after a time astrology became an intimate part of the healing art.

In almost every work on medicine during the Middle Ages the importance of the influence of the stars and constellations is impressed on the physician, and figures representing the internal organs and the signs associated with them are represented in many manuscripts and early printed books. Detailed warnings are given against the treatment of wounds in particular parts of the body according to the sign through which the moon was passing at the time of the injury. This idea is thought to have passed by way of the Gnostics into medieval medicine, the pagan gods being replaced partly by the planets and Zodiacal signs.

Astrologers called the imaginary divisions of the twelve signs the twelve houses, and persons born under a certain sign were supposed to have to some extent its properties and nature. They also believed that the influence of the planets and stars corresponded with the medicinal properties of certain plants which might act for good or evil, if they radiated on corresponding elements in the body of man. They asserted, therefore, that if they knew the influence of a star, the conjunction of the planets, and the qualities of the medicine, they would know

AN ALCHEMIST AND AN ASTROLOGER IN CONSULTATION OVER A PROCESS TO BE CARRIED OUT WHEN THE MOON IS IN THE SIGN OF SCORPIO
From a woodcut
Brunschwig, 1507

what remedies to give to attract such influence as might act beneficially.

Throughout the writings of the early philosophers there is constant reference to the mystic connexion between the seven metals and the seven planets, to which Stephanus of Alexandria adds the seven colours and the seven transformations; consequently, in alchemical symbolism the same sign came to represent the metal and its corresponding planet.

Thus in subsequent years astrology became closely related to

THE INFLUENCE OF ASTROLOGY

the Hermetic Art in its development, and the alchemist came to be regarded as an authority not only on the transmutation of metals, but also on astrology and magic.

The alchemist believed that the planets had the power of maturing metals in the earth and could thus influence his operations, which aimed at their transmutation. He inferred that special forms of matter were more or less directly under the influence of the heavenly bodies, and in time the connexion became fixed in the symbols he employed. The favourable hours for experiments were based on the theory that the seven planets were associated with the seven metals.

The alchemist accepted the astrological doctrine that each planet governed some mineral: the Sun ruled gold, the Moon silver, Mars iron, Venus copper, Saturn lead, Jupiter tin, and Mercury quicksilver. These, according to the alchemists, were the seven bodies.

The four spirits were mercury, sal ammoniac, arsenic, and sulphur, which, says Lyly, a philosopher, "are the fundamental things by one of which the bodies are changed. These bodies seven and spirits four whose total is eleven, being properly calcined, dissolved, coagulated, distilled, and cohobated, are the whole matter of the Stone."

Magic and divination, which were inseparably bound up with astrology, also came to be associated with alchemy. In all these occult sciences the supreme power was believed to be in the stars, and from their mysterious emanations all the metals, precious stones, plants, and herbs derived their special properties.

As late as 1317, in some countries, the alchemists were classed with the astrologers. According to a decree issued by Pope John XXII in that year:

> Alchemies are here prohibited, and those who practise them or procure them being done are punished. They must forfeit to the public treasury for the benefit of the poor as much genuine gold and silver as they have manufactured of the false or adulterated metal. If they have not sufficient means for this the penalty may

be changed to another at the discretion of the judge, and they shall be considered criminals.

Later it was taught that the influence of the heavenly bodies would not be efficient without some intermediate agency to unite them with bodies subjected to them. This agency is light, and by this the heavenly bodies manifest natural effects, and by motion communicate the application of this light, and as all planets receive their light from the sun their influence is varied in proportion to their mutual action.

The variety of colours in the heavenly bodies causes a variety of effects; thus the colour of the sun is almost like gold, and that of the moon nearly white. The sun's colour was endued with an active quality because it proceeds from intension.

The sun and the moon were believed to have greater influence over the human body than all the other heavenly bodies, and to affect mankind in various ways when they entered a certain sign of the Zodiac.

It is evident that astrology, which was probably the most ancient of the occult sciences, had a marked influence on alchemy, in that during the Middle Ages the two so-called arts were constantly practised together.

CHAPTER VI
THE GREEK ALCHEMISTS

OUR knowledge of the Divine Art, later called alchemy, as practised in ancient Greece is chiefly derived from a collection of chemical recipes in some Græco-Egyptian papyri now preserved at Leyden, and a number of manuscripts, notably one at St Mark's Library in Venice, written about the tenth or eleventh century. That the Greeks in Egypt were practising the art about A.D. 290 is evident from the edict issued by the Emperor Diocletian commanding that diligent inquiry should be made "for all the ancient books which treated of the art of making gold and silver," and that such books should be burned and destroyed.

In Europe the Art, which apparently began with the technical operations of metallurgy, had no special name other than the Divine or Sacred Art until the fifth or sixth century of our era. It appears to have originated in the efforts of the workers in metals to imitate gold. They knew its properties and value, and the problem arose as to how might the commoner metals be changed into the one that was the most precious. This idea no doubt deeply impressed these men, for a successful solution of the problem meant not only wealth, but power. Their early conclusions are to be found in one of the papyri at Leyden which is supposed to have been written about A.D. 300, and is believed to be a copy of one of the ancient Egyptian books on the preparation of gold and silver. It contains considerable information on metallurgical operations and records a hundred and one recipes. It appears to have been the formulary of a goldsmith, as directions are given for making alloys for use in the manufacture of vases, images, and cups, with instructions for

ALCHEMY AND ALCHEMISTS

colouring them, together with methods of making gold and silver links. There is also a formula for preparing an amalgam of copper and tin, which was probably intended as an imitation of gold, and a method is described of blackening metals by means of sulphur. These early workers knew the physical characters of the metals, and by making suitable alloys they no doubt found methods of forming one alloy that had the appearance of gold. A number of the recipes have for their aim the production of an alloy which shall be indistinguishable from the real metal, but no idea of transmutation is apparent in the papyrus.

In some of the recipes a "never-ending" material and an "asemon" are mentioned. The latter is believed to have been an alloy of gold and silver, probably the Egyptian electron, as a material which could give the qualities of the precious metals to the baser ones.

In another papyrus allusion is made to the making of imitation precious stones and dyes. Colours played an important part in the operations, the idea being that if a metal could be dyed its nature would be changed and it would become like the metal it was designed to imitate.

The two primary dyes that gave the colours of the precious metals were xanthosis, capable of dyeing yellow, and leucosis, capable of dyeing white. Although they were different in appearance, it was thought that these two dyes were one and the same substance; therefore, it was argued, there must be some more powerful dye, if it could be discovered, that would really change the baser metal into the one it was sought to imitate. In this we have the germ of the idea of transmutation which gradually developed into the conception of the Philosopher's Stone. The greatest secrecy was observed regarding these recipes for making the alloys, which at first were handed down orally from generation to generation by the operators. There were secrets in all trades, and in the early indentures the apprentice was bound not to divulge his master's secrets, a

THE GREEK ALCHEMISTS

command mentioned in one of the Leyden papyri in the words: "He puts out terrible oaths in order that they may be revealed to none."

To one of the later alchemical manuscripts, written about the tenth century, there is added a recipe for making an "excellent Gold Hide," "a sacred secret which ought not to be disclosed to anyone nor given to any prophet."

The first Greek writer who mentions the transmutation of metals is Æneas Baræus, who lived about the fifth century A.D., and wrote a commentary on the works of Theophrastus, but no mention of the Philosopher's Stone is made until we come to the seventh century. The treatises of the Greek alchemists that have come down to us are few and fragmentary, and are chiefly a mixture of magical formulæ, astrological ideas, together with mystical allusions to an earlier philosophy. There is, however, evidence to show that they were acquainted with a number of ores, minerals, and the working of metals.

Zosimus of Panopolis, who flourished about the fifth century of our era, is the first alchemist of whom we have any authentic record. That he was a man of unusual skill in the Hermetic Art may be gathered from references made to him in works of the eighth and ninth centuries by Photius and other historians, who allude to him with veneration. According to Suidas, he wrote twenty-eight books on alchemy, but "most of these were destroyed in the past." In one of his treatises he gives an interesting description of several pieces of apparatus which were employed in the laboratories of his time. The alembic used for distillation, he tells us, was made of glass with a stem of clay or terra-cotta, and another was used for the fixation of mercury. An alembic made with three stems or tubes, which he calls a tribicus, was also used, and he further mentions a still-head, a receiver, and other vessels for using with a furnace. He states in one of his works that, besides the doctrines contained in the hermetic books, there was "the secret knowledge of the priests enveloped in mystery which it was forbidden to reveal," which shows that much of the Greek knowledge of the art was acquired from Egypt.

Zosimus wrote works called *The Great and Divine Art of the Making of Gold and Silver*, *The Book of the Truth of Sophe the Egyptian*, and *The First Book of the Achievements of Zosimus of Thebes*. In the last-named he observes that the kingdom of Egypt was maintained by the art of making gold, and also

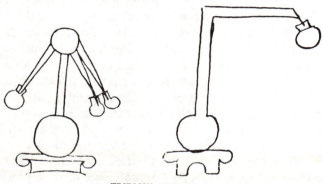

TRIBICUS AND STILL
From a Greek manuscript of the eleventh century

alludes to the ancient stelæ on which, in that country, the secrets of the art were inscribed in symbolic characters. He mentions some of the apparatus used in the Temple of Memphis, and refers to the pneumatic and mechanical works of Archimedes. In the Ninth Book, called *Imouthe*, allusion is made to I-em-hêtep, the son of Ptah, an early Egyptian deity of healing, and he also mentions the book *Chema*, which, according to tradition, was given by the angels to mortals. Other treatises attributed to him are *The Book of Keys* and *The Keys of Magic*, one on the tempering of bronze and iron, one on making glass, and another dealing with lime, which, he says, is "the secret that must not be revealed."

Another of the writers on the art of this period was Sextus Julius Africanus, who is said to have been an Assyrian by birth.

THE GREEK ALCHEMISTS

He was the author of several works on alchemy, medicine, and the natural sciences, as well as on geography and military subjects.

The most important figure of the fifth century in alchemy was Synesius, who in the year 401 was made Bishop of Ptolemais. A man of great intellect, he was in turn physician, astronomer, agriculturist, and diplomat. In the latter capacity he was sent as ambassador on a mission to the Emperor Arcadius at Constantinople. He wrote several philosophical treatises, and also a work on dreams and their interpretation, with which he includes a number of alchemical recipes.

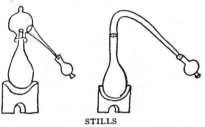

STILLS
From a Greek manuscript of the eleventh century

Among the works attributed to him is one addressed to Dioscorus, priest of the great Serapis of Alexandria, which begins: "Synesius the philosopher greets you." In another treatise, probably apocryphal, called *The Old Book of Dr Synesius, Greek Abbot*, there is a paragraph which says: "Take from them that living silver and you will make it the medicine or quintessence, the imperishable and permanent power, the bond of all elements which contained within itself the spirit which unites all things."

Another alchemical philosopher of whose life but little is known was Olympiodorus, who is the reputed author of a treatise called *Olympiodorus, Philosopher to Petasius, King of Armenia, on the Sacred and Divine Art*. Whether he was the Greek historian of this name who flourished about the year 412 is uncertain. Among his predecessors he mentions Synesius and Mary the Jewess, while he also refers to the Oracles of Apollo and to inscriptions on the Temple of Isis. He relates ancient traditions as to the origin of gold, and says it was produced in the soil of Ethiopia. He refers also to the tomb of

ALCHEMY AND ALCHEMISTS

Osiris and the alchemical symbol of the serpent biting its tail, as well as to the Zodiacal signs, from which it may be inferred that he had travelled in Egypt. He mentions the library of Ptolemy at Alexandria as though he had visited it, and quotes the opinions of the Greek philosophers of the Ionic school. Like other early writers, he comments on the symbolism used by the alchemists to ensure secrecy and on the obscurity of the language they employed. "The ancients," he says,

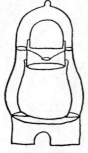

APPARATUS FOR DIGESTION
From a Greek manuscript of the eleventh century at Leyden

> have the custom of hiding the truth and of veiling and obscuring by allegories things which are clear and evident to all the world. They admit three tinctures. The first disappears rapidly [is volatilized] like sulphur and arsenic. The second disappears slowly, like sulphurous materials, and the third will not disappear at all. These are metals, stones, and earths.
>
> The first tincture, which is made with arsenic, contains white copper. Arsenic is a kind of sulphur which is readily volatilized; all that resembles arsenic is volatilized by fire, and is called sulphurous matter.

Olympiodorus gives particulars of a second tincture, which was used in the making of emeralds, the formula for which was as follows:

> Take two ounces of fine crystal and half an ounce of calcined copper. First of all prepare the crystal by the action of fire. Place it in pure water, wipe it, crush the substances in a mortar, and fuse them together at an even temperature.

Kopp, who made a very thorough research in connexion with the authors of the early Greek manuscripts, lays importance on one who passed under the name of Democritus. By some critics he was believed to be identical with the Greek philosopher of the same name, but there is no proof of this. Democritus of Abdera travelled and studied in Egypt. He became acquainted with the secret lore of that country, and wrote

THE GREEK ALCHEMISTS

several works about the secrets of nature. Some of his writings appear to relate to the methods of the Egyptian alchemists, so that it is possible that those writings represent some part of the lore which the elder Democritus had acquired.

The study of alchemy in Greece appears to have reached its highest point about the fifth or sixth century and during the reign of Theodosius I and his successors. The Temple of Serapis was the centre of Greek civilization in Alexandria, and it is there probably that alchemy was chiefly studied. Its destruction by command of Theodosius was a calamity to the history of both science and art. The Serapeum of Memphis and the Temple of Ptah, which are thought to have been the seats of the study of alchemy in Egypt, were destroyed about the same time as the Alexandrian schools of learning.

Mention should here be made of the part the Jews played in the fusion of the Oriental doctrines of religion and science with those of Greece in the early years of the Christian era. For a time the Alexandrine Jews took the lead in the arts and in philosophy, and there is evidence of the connexion of the Jews with alchemy both in the Leyden papyri and in the early Greek manuscripts. In the papyrus of Reuben there is an alchemical recipe attributed to Hosea, King of Israel, and in others references are made to Abraham, Isaac, and Jacob, while other Jewish names are found in the alchemical writings of the third century. There is also a papyrus dealing with magic and astrology entitled *The Holy Book*, or the *Eighth Monad Book of Moses*, the *Key of Moses*, and the *Secret Book of Moses*. The Jews probably acquired a knowledge of the Divine Art from the Egyptians, for Zosimus tells us that "the Sacred Art of the Egyptians, and their control of gold which resulted from it, was only revealed to the Jews by a fraud, and they made it known throughout the rest of the world."

Mary the Jewess, a mysterious figure, whose name appears in some of the early manuscripts, is said to have written several treatises on alchemy in one of which is said: "Touch not the

Philosopher's Stone with thine hands; thou art not of our race, thou art not of the race of Abraham." Her name is perpetuated in the Balneum Mariæ (Bath of Mary), which, according to tradition, she originated.

The Gnostics, that strange sect among the early Christians who claimed to have a superior knowledge of spiritual matters and interpreted the sacred writings by a mystical philosophy, are also said to have been skilled in alchemy. Gnostic names are mentioned by the early Greek writers on alchemy, and, according to Reuben, the recipes in the Leyden papyri are founded on Gnostic ideas, and are attributed to Marcus, the Gnostic prophet, who is said to have been born in Palestine. The serpent with its tail in its mouth is often found engraved on stones with magical symbols of the Gnostic period, and this, as well as the concentric circles, was used as an emblem of the Philosopher's Egg, the sign of the Universe, and of Alchemy. The serpent was considered both good and evil, as instanced in the Egyptian Apôp or Apophis, the snake-giant, and as symbolic of the powers of darkness and their war against the Sun.

The Stoics also, in their endeavour to prove that the whole Cosmos was permeated by the divine, became supporters and defenders of magical, alchemical, and astrological ideas, and there seems little doubt that they had a considerable influence on the development of the art in Greece.

CHAPTER VII
ALCHEMY IN CHINA AND INDIA

RECENT research has shown that as far back as the fourth century before the Christian era certain Chinese philosophers sought the secret of prolonging human life beyond its usual span, with immortality as their goal. The first conception was to achieve this by physical means such as gymnastics, proper breathing, and mental training. Further, it was to be attained by the proper use of food and the employment of medicines and of certain substances containing vitalizing qualities. Such substances were supposed to be animated by a vital spirit, and relative to this idea there is a very ancient legend. On the island of Yong Chou there was said to be a mountain, composed of jade, about 10,000 feet high, from which issued a spring with a sweet taste resembling that of wine. It was called Jade-wine Spring, and those who drank several pints of it would immediately fall into a state of intoxication, after which immortality was assured.[1]

The cult of Taoism also was associated with Chinese alchemy from early times, and the monks of the sect, who were followers of the philosopher Lao-Tzŭ, are said to have practised magic.

According to the Chinese encyclopædia called *Pei-ouen-yun-fou*, the first who purified the *Tan* was called Ko-hong. This man lived during the dynasty of Ou—that is, between A.D. 222 and 277. *Tan* was a sacred and technical expression which meant the search for the secret for the transmutation of metals, and so we have evidence of the dual quest in China—that is, to discover how to prolong life and how to transmute metals, although the former appears to have been the first object of their search.

[1] D. S. Johnson, *Study of Chinese Alchemy*.

ALCHEMY AND ALCHEMISTS

Among the animal, vegetable, and mineral vitalizing substances sought for by the Taoists were cinnabar, gold, silver, and jade, and these were held in the highest estimation. Cinnabar became a favourite ingredient in life—prolonging elixirs by virtue of its producing, when heated, the 'living' metal, mercury. It was claimed to be capable of converting other metals into gold, as well as of being able to prolong life for an indefinite period. Of the vegetable substances with vitalizing properties there were the pine-tree, peach, and the *chih* plant, known as the 'divine herb.' The products of the pine-tree, if refined and eaten, were believed to prolong life, and the peach also was considered to have the same property. The crane, fowl, and tortoise were regarded as possessing life-giving virtues, and the eggs of the crane formed an ingredient in many of the recipes for making the elixirs.

Ko-hong, in a work said to have been written in the third century, says:

> When vegetable matter is burnt it is destroyed, but when *iansha* [cinnabar] is so treated it produces mercury, and after passing through other changes it returns to its original form. It differs widely therefore from a vegetable substance, and hence it has the power of making a man live for ever and raising him to the rank of the genii.
> He who knows the doctrine is a man far above common men.

It is interesting to note that China appears to have had an extensive literature dealing with alchemy prior to the period when the art was studied in Europe; when the alchemists of the West were seeking the Philosopher's Stone the Chinese claimed to have discovered it in the form of cinnabar.

The similarity of the ideas and doctrines of the East and the West is striking, but it must be remembered that there was considerable intercourse between China, India, and Persia, and that after the Mohammedan conquest of the latter country embassies from Persia and Arabia, and even from the Greeks in Constantinople, visited the Court of the Chinese Emperor in

ALCHEMY IN CHINA AND INDIA

Shansi; Arab traders settled in China, and there was frequent intercourse by sea between China and the Persian Gulf.

There is a curious legend concerning a Chinese alchemist and philosopher named Wei-po-Yang, who flourished in the second century and wrote a treatise on a preparation called the Pills of Immortality, which after prolonged study he is said to have ultimately succeeded in making. He administered one of his pills as an experiment to a dog, but unfortunately the animal speedily succumbed to the effects, whereupon he swallowed a pill himself and he also died. His elder brother, who still had a firm faith in the pills, now took a dose, but he too fell down dead. These terrible results not unnaturally shook the confidence of a younger brother, and he resolved not to risk his life and went off to make arrangements for the burials. Much to his amazement, he found on his return that all the victims had completely recovered. Thus Wei-po-Yang was enrolled among the immortals.

The search for the secret of how to prolong life, which appears to have been the first phase in Chinese alchemy, seems in the early centuries of our era to have developed into a quest for an agent which would be both capable of producing an elixir of life and a means of transmuting metals. The seekers after this knowledge had a curious resemblance to those who practised alchemy in the West. There were those who had a genuine and unselfish interest in the science—largely recluses or anchorites who pursued their studies in the solitudes of the mountains; there were others who, seeking personal glory and wealth, frequented the Imperial Courts. According to a Chinese writer, the former class sought the following: "Eight precious things —cinnabar, orpiment, realgar, sulphur, saltpetre, ammonia, empty green [an ore of cobalt], and mother-of-clouds [a variety of mica]."

The philosophers considered that base metals might be transmuted into the precious ones by the dual method of eliminating the more material qualities in their composition and of augmenting the more spiritual qualities. Visible changes in either colour, form, or ductility would be considered as actual proof that the process of refinement or transmutation was taking place. Whiteness, they contended, was the property of lead, but if it was caused to become red the lead was changed to cinnabar, as redness was the property of cinnabar. They thought, therefore, that by a process of refinement the baser metals might be artificially transmuted to the more precious metals, providing only that the proper refining medium or parent substance might be discovered.

The Chinese alchemists believed that transmutation was an inherent feature in the process of growth, for, according to Liu An (122 B.C.), "gold grows in the earth by a slow process and is evolved from the immaterial principle underlying the universe, passing from one form to another up to silver, and then from silver to gold." Here we have the same doctrine as taught centuries later by alchemists in the West.

The first recorded attempt to transmute metals by artificial means alluded to in Chinese literature is that of Li Shao-chün, the master alchemist at the Imperial Court during the reign of the Emperor Wu Ti, who flourished from 140 to 86 B.C. He states:

> I know how cinnabar transforms its nature and passes into yellow gold. I can rein the flying dragon and visit the extremities of the earth. I can bestride the hoary crane and soar above the nine degrees of heaven. If you will make sacrifice to the furnace you will be able to transmute cinnabar to gold.

Some writers say that the mysterious Tan Stone was one of the agents necessary to effect transmutation, and that a small quantity was always placed in the crucibles employed in the process. Others refer to the agent as the exoteric drug or the golden drug, but the majority agree that some form of mercury formed the basis of it.

Hanbury, who lived in China in the last century, states that he believed that cinnabar itself constituted the agent, as mercury

ALCHEMY IN CHINA AND INDIA

was regarded by the Chinese as the 'Soul of Metals' or the 'Living Metal,' and was likened to the human soul. According to *The Book of the Immortals*, the essence of cinnabar produces gold. "Lead is the mother of silver, and mercury is the child of cinnabar [mercury bisulphide]."

In the biography of Lü Tsu, a Chinese alchemist who is said to have lived in the eighth century A.D., there is a story told of a person named Wu Ta-wen, who lived in the city of Chengtu. He held an official position and was a man of great knowledge. At one time he was in the service of a Taoist magician named Li Ken, whom he one day saw heating lead and tin together over a fire. The magician took a small quantity of some drug about the size of a bean, threw it into the crucible, and stirred the mixture with an iron spoon. Upon cooling it immediately changed to silver. Wu Ta-wen obtained the secret formula, and whenever he wished to make use of it he abstained from eating meat for a hundred days, and then retired from the public gaze, preferably to the summit of some lofty mountain, there, in complete secrecy or in the presence of a few devotees of undoubted trustworthiness, to carry out the process.

The Chinese alchemists, like the adepts in the West, employed symbols and allegories in their writings in order to ensure the secrecy of their formulæ and processes. Thus we have allusions to Yin, the principle of cold and humidity, as the white tiger, and to Yang, the principle of heat and dryness, as the green dragon. Realgar was known as the masculine yellow, orpiment as the feminine yellow, and mother-of-pearl as the Cloud Mother. Vermilion was called the Fairy Lady, the sun the Golden Crow, and the moon the Golden Mirror. It is apparent from the early records that alchemy was practised in China centuries before the art was known in Europe, for, according to our present knowledge, it did not make its appearance in the West until about the fifth or sixth century of the Christian era. In both systems there was the twofold quest for the prolongation of life and the accession of wealth, but in the East the former was the dominant feature, while the chief object of the Western alchemists was the secret of transmuting the baser metals into gold.

In India the earliest allusions to alchemical ideas appear in the *Atharva Veda*, where mention is made of the gold which is born from fire: "The immortal they bestowed upon the mortals. The gold [endowed by] the sun with beautiful colour which the men of yore rich in descendants did desire. Long-lived becomes he who wears it."

From very early times in India gold was regarded as the Elixir of Life, and lead was looked upon as the dispeller of sorcery. Dr Ray, an authority on the subject, attributes the rise of Hindu alchemy to the *Soma rasa* plant, which was an object of admiration to the Vedic worshippers, for the juice was regarded as the stimulant which conferred immortality upon the gods. By other authorities alchemy is said to have had its origin in the Tantras, part of the series of sacred books known as the Fifth Veda, supposed to date from the sixth or seventh century of our era. Before the eleventh century, however, records are scanty. Albērūnī, the historian, says:

> Until that period the Hindus did not pay particular attention to it, but many intelligent people are entirely given to alchemy. The adepts in this art try to keep it concealed, and shrink back from intercourse with those who do not belong to them. They have a science similar to alchemy which they call Rasāyana, an art which is restricted to certain operations, drugs, and compound medicines. Its principles restore the health of those who were ill beyond hope and give back youth to fading old age.

Among the Tantras of the eleventh and twelfth centuries many allusions are to be found concerning quicksilver, which was supposed to be capable of giving a divine body. The work called *Rasārnava* states: "It is mercury alone that can make the body undecaying and immortal." It was the supreme medicament, and its study was regarded as a science in itself. During the Tantric period, with its system of the "Philosophy of Mercury," a great amount of chemical information was accumulated

ALCHEMY IN CHINA AND INDIA

mainly concerning the search for the Elixir of Life, and the knowledge of the Hindus on the subject appears to have been equal to that obtaining at the same period in Europe.

In the Vedic period gold was described as the yellow metal, silver as the white metal, iron as the black metal, and copper as the red metal, while, in the time of Susruta, who is called the father of Hindu surgery, tin and lead were also known. Of alkaline substances sodium carbonate (natron), potassium carbonate, and borax were employed, and acids were obtained from citrons, lemons, and other vegetable sources. Treacle, clarified butter, and honey were used as solvents, and borax as a flux for metals.

Nāgārjuna, the philosopher, was looked upon as the originator of the processes of distillation and calcination, and it is claimed that arsenic, zinc, and iron were used in the tenth century, while the processes for the calcination of tin, copper, and lead were known about the same period. Metallic preparations were employed from about the fourth century, and were generally prepared by the process of 'killing' the metals, a method frequently employed in India. This meant depriving a metal of its characteristic physical properties, such as its colour and lustre. Both gold and silver were treated in this way and were reduced to a fine state of division. The methods employed for carrying this out varied, both heat and trituration being used, and sometimes the juices of certain plants were added during the process. The Hindu alchemists took especial care that the operation should be carried out effectively, and in the case of mercury it was directed "to subject it to a gentle heat for three hours."

In *Rasārnava*, which is supposed to have been written about the twelfth century A.D., some interesting particulars are given with reference to the equipment of an alchemist's laboratory. It is recommended that the laboratory should be erected in a region which abounds with medicinal herbs and wells. The furnaces should be arranged on the south-east side and instruments on the south-west. Washing operations should be carried out on the west side of the building, and drying in the north-west. It should be furnished with bellows, mortars, pestles, sieves, earth for crucibles, charcoal, conch-shells, iron pans, and retorts made of glass, earth, and iron. The crucibles should be composed of black, red, yellow, or white earth, and apparatus for distillation, sublimation, and for killing metals installed. It is further recommended that lutes should be made of lime, raw sugar, rust of iron, and buffalo's milk, and cow-dung cakes should be used for fuel.

Among the writings of the Hindu alchemists in the seventeenth century many recipes are to be found for making artificial precious stones, an art that was also practised by the early alchemists in Europe. Some of the recipes contained in a manuscript that was translated about a century ago, which are probably of greater antiquity, show a considerable knowledge of chemistry, as may be judged from the following examples:

The Way to make Artificial Diamonds

To counterfeit diamonds so as to endure the fire and harden them, take of good natural crystal and reduce to subtle powder. Fill a pot with it, set it in a glass-house furnace for twelve hours to be melted and purified, then drop the melted matter into cold water, then dry it and reduce it again to powder and add to it its weight of fine salts of tartar. Mix these two powders well and make little pills of them with common water. Then wipe these pills and put them into an earthen pot on a strong fire, there to grow red for twelve hours without melting. Then put them into a pot in a glass-house furnace, where let them stand two days to be well melted and purified. Then set the matter for twelve hours in the annealing furnace to cool gradually. Afterward break the crucible and you will have a fine material for diamonds, which must be cut and polished at the wheel.

To make Emeralds

Take calcined brass 3 ounces in powder, which recalcine with oil and a weaker fire for four days. Use this to colour glass or add to the crystal *crocus martis* and brass, twice calcined.

ALCHEMY IN CHINA AND INDIA

To make a Green Emerald

Take glass made without manganese; to this add scales of copper, thrice calcined, and scales of iron from a smith's forge, both well washed, pounded fine, and sifted. These scales of iron serve instead of *crocus martis*.

To make a Ruby

Take oxide of manganese and strass.

To make a Topaz

Take glass of antimony, purple of Cassius, oxide of iron, and white strass, and fuse together.

The idea of transmutation is evidenced in the following process for changing white sapphires into true diamonds:

The white sapphire, being fine and fixed, is only imperfect by reason of its wanting colour and hardness, which may be remedied by art and be made to surpass Nature because she only would have made it a perfect sapphire, but art can turn it into a true diamond. It is only fire can work this effect. Thus take very fine sand, wash it in several waters till the water becomes clear, then dry it. Fill a crucible half full with this sand; then put in your sapphire, and fill it up with the same sand.

Then cover your crucible with a cover of the same earth or another crucible. Lute the whole with a good lute and lay it on an inch thick and let it lie in the shade. Being dry, set it in a glass-house furnace, approaching it nearer to the fire by degrees and leaving it twelve hours in the same heat. Then withdraw it little by little for the space of six hours and let it cool. Break it and you will find therein your sapphire, which will have all the qualities of a fine diamond. Polish it again at the wheel and work it.

In a book written by Sānāq, an Indian, chemical analysis is foreshadowed in a method described for detecting poison in food or drink. He observes:

The vapour emitted by poisoned food has the colour of the throat of the peacock. . . . When the food is thrown into the fire it rises high in the air. The fire makes a crackling sound as when salt deflagrates. The smoke has the smell of a burnt corpse.

Poisoned drinks, buttermilk, and thin milk leave a light blue to yellow line.

It is probable that the knowledge of alchemy in India was acquired from the Chinese, for as early as the second century before the Christian era the Emperor Wu Ti (140–86 B.C.) sent an embassy into Central Asia, which proved an important factor in spreading and facilitating intercourse between China and the West.

CHAPTER VIII
THE ARABIAN ALCHEMISTS

WITH the conquest of Egypt, Syria, and Persia by the Arabs and the rise of Islam the centre of scientific learning changed. The Arabians were eager seekers after knowledge, and became the most cultivated people in the world. The first impulse given to the desire for knowledge of the wisdom of the Greeks came from the Umayyad Prince Khálid, who had a passion for alchemy. According to the *Fihrist*, the oldest and best existing source of our knowledge on the subject, this prince assembled the Greek philosophers in Egypt about the eighth century A.D., and commanded them to translate Greek and Egyptian books on this subject into Arabic. These, he says, "were the first translations made in Islam from one language to another." The Arabs thus rekindled the ancient lamp of science that had grown dim in Europe. They established universities, like that at Baghdad, which became great centres of learning and later had so important an influence on the spread of scientific knowledge.

Many of their students became celebrated as physicians and alchemists, and foremost among these pioneers was Jábir-ibn-Hayyán, who became famous in medieval Europe under the name of Geber. He flourished in the eighth century, a period when the great stream of Greek and other ancient learning began to pour into the Mohammedan world and to reclothe itself in Arabian dress.

Jábir is said to have been born about A.D. 702 and to have been either a native of Mesopotamia or a Greek who afterward embraced Mohammedanism. Little is known of his life beyond the fact that he was associated with Prince Khálid and that he devoted himself enthusiastically to the study of alchemy. According to one historian, he practised medicine at the Court of Haroun-al-Raschid and lived for some time at Kufa, where, many years after his death, the remains of his laboratory were discovered when some houses in that city were being demolished. He was the reputed author of a great number of works, but many of those which have come down to us in Latin are said to be spurious, being the work of medieval alchemists who sought, by taking advantage of the prestige attaching to his name, to give authority and currency to their own writings.

The Arabic originals of Jábir's works are rare, and the British Museum possesses but two manuscripts, said to be the only copies in existence, of his book called *The Great Book of Properties*. Other works attributed to him are *The Book of the Divine Science*, *The Book of Definitions*, and the *One Hundred and Twelve*.[1] His works show that he preferred the practical work of the laboratory to theorizing, and they express his views that chemistry is that branch of natural science which investigates the properties and generation of minerals and substances obtained from plants and animals. He advises that the alchemist should know the reason for performing each operation, and that those which are profitless should be avoided. He should choose trusty friends and select a secluded place to carry on his work. He also urges the taking of time in carrying out experiments, together with patience, reticence, and perseverance on the part of the operator, whom he finally adjures not to be deceived by appearances and not to bring his experiments to too hasty a conclusion. He advocated careful observation and the rejection of any assertion that could not be supported by proofs.

He considered that all metals were compounds of sulphur, mercury, and arsenic, the differences between them depending on the preparation and degree of purity of each body. Gold

---

[1] Recent historians assign some of the books attributed to him to a period between the ninth and tenth centuries.

THE ARABIAN ALCHEMISTS

in particular he believed to be composed of purified mercury mixed with a small quantity of pure sulphur. He thought that metals having common constituents could be transmuted one to another, and that in the performance of this operation the alchemist was only doing what Nature herself had performed. He contended that, as the metals were but different mixtures of sulphur and mercury, the precious metals being richer in mercury than in sulphur, the transmutation of lead or copper into gold or silver meant the withdrawal of sulphur from and the addition of mercury to them.

Jábir was apparently familiar with the seven metals, and knew that some of them could be converted into earthy powders by burning them in air as well as by other processes. It was by the prolonged oxidation and calcination of a metal, the reducing of it to a natural state, and the repetition of these operations, that its active properties could be discovered—an idea also held by alchemists in the Far East.

Jábir knew of the oxides of copper, iron, and mercury, and also of the yellow and red oxides of lead. He is credited with the discovery of arsenious oxide (white arsenic), and refers to its power of whitening copper. He describes three varieties of alum, green vitriol, sal ammoniac, borax, saltpetre, and a method of making common salt. He is also associated with the introduction of corrosive sublimate, silver nitrate, red oxide of mercury, and terchloride of gold in solution. He refers to sulphuric and nitric acids, and made acetic acid from the distillation of vinegar, while he also knew of the alkaline carbonates, such as those of potassium and sodium.

The chief operations and processes he employed were sublimation, solution, filtration, and crystallization, and he also describes digestion at various degrees of heat, together with the construction of furnaces. He observes respecting sublimation:

> Mercury, sal ammoniac, sulphur, and arsenic are substances that are capable of sublimation, but marcasite and pyrites on being heated in an earthen distilling apparatus give sulphur which leaves

a residuum, which sediment oxidizes little by little under the influence of air in the apparatus, and part of the product sublimates at a higher temperature, furnishing white or coloured metallic oxides.

He describes the operation of cupellation, or of assaying or purifying gold, by heating it with lead in a porous crucible made of pounded bone-ash, an operation which is probably the earliest known in metallurgical chemistry and which is carried on at the present day. He further emphasizes the importance of distillation to the alchemist, and describes the ordinary methods by ascent and by descent.

Jábir did not neglect the practical application of science to other arts, and refers to a method of preparing steel and refining other metals, while he also mentions the use of manganese dioxide in glass-making. He gives recipes for making an illuminating ink from golden marcasite to be used in place of real gold for dyeing cloth and leather, and for varnishes for protecting iron.

These are but a few of the achievements attributed to Jábir-ibn-Hayyán, that remarkable pioneer in science who may be said to have laid the foundations of chemistry in the eighth century.

In the century following another great figure arose in the person of Abu-Bakr-Muhammad-ibn-Zakariyya-al-Razi, who became commonly known as Rhazes. The date of his birth is unknown, but he is said to have died between the years 903 and 923. He is believed to have been born at Ray, near Tehran, in Persia, and as a youth was a musician and skilful player on the lute. Later he became a student of medicine, and was so successful that he was made chief physician at the hospital at Ray and afterward physician-in-chief to the great hospital at Baghdad. It is said that when he was consulted about the building of this institution and asked to suggest a suitable site for it, in order to select a healthy locality, he caused pieces of meat to be hung in different quarters of the city, and chose the place where they were slowest in showing signs of decomposition.

THE ARABIAN ALCHEMISTS

His works on medicine were many, but the one by which he will always be remembered is his treatise on smallpox and measles, two diseases which he was the first to describe. This work ranks in the history of epidemiology as the earliest monograph upon these diseases.

He was the first also to make an attempt to classify drugs and chemical substances, and to group them as animal, vegetable, and mineral. The last-named he again divided into six classes—spirits, bodies, stones, vitriols, boraces, and salts. He regarded sulphur, arsenic sulphide, sal ammoniac, and mercury as the spirits; tin, lead, and iron as metals or metallic bodies; magnesia, marcasite, tuttia (impure oxide of zinc), lapis lazuli, alum, antimony sulphide, talc, gypsum, and glass as stones. The vitriols he classified according to their colour—black, white, red, green, and yellow. The boraces included borax, natron, bone-ash, and tinkar, and the salts comprised cooking salt, sweet salt, bitter salt, bituminous salt, calcined salt, quali (crude soda carbonate from ashes of maritime plants), and salt of ashes (crude potassium carbonate from ashes of land plants).

Judging from his works on alchemy, Rhazes did not regard the transmutation of metals as its chief aim, but he emphasizes the value of a knowledge of chemical substances as applied to medicine. He throws a light on the antiquity of certain pieces of apparatus in a list he gives of appliances he employed in his laboratory. This includes furnaces, crucibles, a descensory, tongs, ladle, shears, pestles, cauldrons, retorts, receivers, aludels, alembics, ovens, glass flasks, basins, and a flat stone for pounding or for use with a roller for pulverization.

Rhazes is said to have become blind owing to cataract during the latter period of his life. His loss of sight is, however, said by some to have been due to close application to his alchemical work. Others advance the following explanation. An offer of a large reward was made to him by a great personage if he would apply his knowledge of alchemy to the actual reproduction of gold. Rhazes declined the offer, whereupon the great man lost his temper, accused him of fraud, and struck him so heavily on the head that his eyes were affected, and eventually he lost his sight.

He left several works on alchemy, in one of which he mentions *aqua vitæ*, which was probably a variety of wine.

Many of the Arabian physicians practised alchemy, and among them was Avicenna, who became famous in the eleventh century. He considered that each of the seven metals was a distinct species of one genus, just as the plant genus includes different species, but that it was not possible to convert one into another.

Much of the learning and literature of the Alexandrian schools was preserved by Syrian scholars who took refuge in Persia and there translated into Syriac a number of Greek works on alchemy. Some of these were again translated into Arabic, while the Arabs themselves translated some of the chief works of the Greek philosophers into their own language. It is evident that they also gathered something from India and the Orient, probably through the medium of Baghdad, and so became acquainted with Chinese alchemy.

About the end of the tenth century Abu Mansur Muwaffah, a Persian alchemist, distinguished between sodium carbonate (natron) and potassium carbonate, stating that the latter is obtained from the ashes of certain plants and is a white solid which deliquesces with a caustic taste. He accurately describes antimony, and refers to arsenious oxide as a powerful white powder and to copper oxide as having poisonous properties. He also mentions that gypsum when heated becomes a kind of lime which, when mixed with the white of egg, forms a plaster useful in the treatment of fractures of bones, thus alluding for the first time to the use of plaster of Paris as an aid to surgery.

About the eighth century the Arabs carried a knowledge of alchemy into the southern part of Western Europe, and a little group of philosophers arose in Spain. Cordova was their centre of learning, and later, in the tenth century, it became famous and attracted many students. Among them was Maslaman-al-

THE ARABIAN ALCHEMISTS

Majriti, who was born at Cordova and who, after obtaining further knowledge of alchemy in Arabia, returned to the land of his birth. He is said to have been the author of a book entitled *Rutbatu'l-Hakim* (*The Sage's Step*), supposed to have been written about the middle of the eleventh century, in which he

SYMBOLIC FIGURES　　　　　　　　STILLS
From a twelfth-century Arab manuscript on alchemy

mentions the training necessary for the student who would succeed as an alchemist, and first recommends a thorough mathematical training in natural science, to be followed by practical experience in the laboratory. He states that "Alchemists must try to follow Nature, whose servants indeed they are. Like the physician, they must diagnose the disease and administer the remedy, but it is Nature who acts."

ALCHEMY AND ALCHEMISTS

He gives the first clear description of mercuric oxide and how it should be prepared:

> I took quivering mercury free from impurity and placed it in a glass vessel like an egg. This I put inside another vessel like a cooking pot, and set the whole apparatus over a gentle fire at such a degree of heat that I could bear the hand upon it. Then I would heat it for forty days, after which I opened it. I found the mercury absolutely converted into a red powder, soft to the touch, the weight remaining the same as it was originally.

Among the Syriac manuscripts still extant is one entitled *The Doctrine of Democritus*, which was translated from the Greek between the seventh and eighth centuries. It contains recipes for the preparation of gold and the Philosopher's Stone, while it also mentions sulphur, antimony, and arsenic, and gives a key to the symbols used in the work. It is of particular interest on account of the drawings it contains of various pieces of apparatus employed at the time.

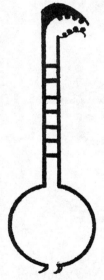

A GRADUATED RECIPIENT
From a twelfth-century Arab manuscript on alchemy

The Moors acquired some knowledge of alchemy from Arabia as early as the twelfth century, and a treatise on the art entitled *The Keys of Providence and the Secrets of Science* was written by Ismail ibn Lhocine Toughrai, who was head vizier to the Emir Messaoud ibn Mohammed about that period. This work is also interesting on account of the drawings which it contains representing some of the apparatus employed.

Before leaving the Arabian alchemists mention should be made of Abu'l-Quasin Muhammad ibn Ahmad-al-Iraqi, who settled in Mesopotamia toward the end of the thirteenth century. He was the author of a work entitled *Knowledge acquired concerning the Cultivation of Gold*, in which he claims that fire alone

THE ARABIAN ALCHEMISTS

is not sufficient to effect transmutation of a metal, and that it is necessary to add some substance such as white or red elixir to accomplish the desired effect. This shows that the conception of the Philosopher's Stone had penetrated to the Near East and was becoming general.

To the Arabs we owe the introduction of alchemy into Western Europe, and that they contributed largely to the body of scientific doctrine they inherited from the Greeks is generally recognized. They freed it from a good deal of the mystery with which it was surrounded, and by their labours discovered many mineral as well as vegetable substances which have proved of the greatest value.

CHAPTER IX

THE PHILOSOPHER'S STONE AND THE ELIXIR OF LIFE

THE lure of the Philosopher's Stone and the quest for the Elixir of Life provide the subject of many of the early romances, and much has been written not only of the legends of how the old have been made young again, but also of the stories of the search for the long-sought-for agent that was capable of transmuting the baser metals into gold, an adventure in which princes and philosophers shared alike.

For ages men of great intelligence and even genius wasted both their lives and their fortunes in the pursuit of this chimera, and in later times the lure of this wonderful substance that was to bring its possessor wealth and power led to both fraud and trickery.

As already stated, we have no mention of the subject until the seventh century, when about 620 Stephanus of Alexandria wrote of "the elements that grow and are transmuted, because it is their qualities, not their substances, which are contrary."

Since the twelfth century claimants to the discovery have been numerous, and many descriptions are recorded of the various processes they have employed for obtaining the coveted medium.

To understand the composition of the agent they sought the seekers had first to learn that it must have the character of sulphur and mercury, which implied certain qualities and not the actual substances now known by these names. Jábir describes sulphur as a "fatness of the earth by temperate decoction in the mine of the earth thickened until it be hardened and made dry."

THE PHILOSOPHER'S STONE

The term 'sulphur' was also used to describe anything combustible, and was sometimes called 'the house of the spirit,' 'elementary fire,' and 'Adamic earth'; and sulphur was regarded as the male element. Mercury, on the other hand, was believed to be the elementary form of all things fusible; it represented lustre, ductility, and malleability.

This sulphur-mercury doctrine was accepted by most alchemists until about the twelfth century, when the theory was extended by the addition of a third elementary principle, to which the name 'salt' was given. It was believed to be a basic principle in the metals which gave to them solidity and resistance to fire. Mercury was considered to be the connecting link between the spirit and the body, and the element on which depended the blood and life.

Many believed that by bringing the elements sulphur, mercury, and 'salt' together they would thus obtain the Philosopher's Stone, one of the names given to the agent required to effect transmutation, and for this mysterious medium a continuous search was made.

The French chemist Lémery says:

> It is this they call the 'Great Work.' . . . Some make a mixture of metals with substances proper to refine them and calcine them a long time in a strong fire to arrive at their perfection. Others seek for a seed of gold in gold itself, and believe they shall find it there as they do the seed of a vegetable in a vegetable. Consequently, to accomplish this they endeavour to open gold by dissolvents. Others look for the seed of gold in the minerals, as in antimony, where they pretend there is a sulphur and a mercury, like to that of gold. Others imagine that they can catch the seed of gold by fixing the rays of the sun after a certain manner, for they, as several astrologers, look upon it as a thing incontestable that the sun is gold melted in the centre of the world.

The so-called Philosopher's Stone was also termed 'the Essence,' 'the Stone of the Wise,' 'the Magisterium,' 'Magnum Opus,' 'the Quintessence,' and 'the Universal Essence.' But, by whatever name it might be called, this object of the alchemist's dreams and striving was regarded as the one perfect thing, and by it

> all infirmities might be cured, human life prolonged to its utmost limits, and mankind preserved in health and strength of body and mind, clearness and vigour. All wounds are healed by it without difficulty, and it is the best and surest remedy against poisons.

The various colours of the metals, and the changes which took place when they were melted with others, were deemed of great importance by the alchemists. In imparting the colour of a noble metal to a base one much was believed to have been attained. The principal operations bear a strong resemblance to the process of dyeing cloth, and the old designation of *tincturæ*, which was applied to the media by which this transformation was effected, gives expression to the idea of a dyeing operation.

The various stages in the perfecting of the 'Stone' are thus described in a work entitled *The Open Entrance*:

> The beginning of the heating of gold with mercury is likened to the king stripping off his golden garments and descending into the fountain. This is the regimen of Mercury. As the heating is continued, all becomes black; this is the regimen of Saturn. Then is noticed a plan of many colours; this is the regimen of Jupiter. About the end of the fourth month you will see the sign of the waxing Moon, and all becomes white; this is the regimen of the Moon. The white colour gives place to purple and green, and you are now in the regimen of Venus. After that, appear all the colours of the rainbow or of a peacock's tail; this is the regimen of Mars. Finally, the colour becomes orange and golden; this is the regimen of the Sun.

Of the mysterious and elusive Stone we have various descriptions left by alchemists who claimed to have discovered the secret, or by others who declared that they had both seen and handled it.

"It is called a stone," says one,

> not because it is like a stone, but only because by virtue of its fixed nature and that it resists the action of fire as successfully as any stone. Its appearance is that of a very fine powder, impalpable

SYMBOLIC REPRESENTATION OF THE BIRTH OF
THE PHILOSOPHER'S STONE
Triomphe Hermetique (1689)

TITLE-PAGE FROM "BASILICA CHYMICA," BY CROLLIUS (1612)
It depicts the "great masters and symbols of alchemy."

THE PHILOSOPHER'S STONE

to the touch, sweet to the taste, fragrant to the smell, in potency a most penetrative spirit, apparently dry and yet unctuous and easily capable of tingeing a plate of metal. If we say its nature is spiritual it would be no more than the truth; if we describe it as corporeal the expression would be equally correct.

It is described as being of various colours, sometimes as a red, white, or black powder, or it may be yellow, blue, or green. Raymond Lully calls it "carbunculus," while Paracelsus declares it to be a solid body like a ruby, transparent and flexible. Beregard says it is "the colour of a wild poppy, with the smell of heated sea-salt," and van Helmont describes it as being "yellow, the colour of saffron, in the form of a heavy powder, with a brilliancy like glass." Helvetius likewise describes it as being yellow and the colour of sulphur, but it is most frequently referred to as the red or white stone. One writer declares that, "although it has concrete form, its working is spiritual." He describes it as of a bluish-grey or green colour. "It is of great weight and is small. It is a certain heavenly, spiritual, penetrative, and fixed substance, which brings all metals to perfection of gold or silver, and that by natural methods." Another alchemist says, "Its colour is sable with intermixed argent which marks the sable fields with veins of glittering argent." The quantity of base metal that could be converted into gold by any given quantity of the Stone evidently varied. Arnaldus de Villa Nova declared that the Stone could convert a hundred times its own weight into gold. Isaac of Holland places its conversion power at a million times, while Roger Bacon states that the Stone could convert a hundred thousand times its own weight of a base metal. It was sometimes called the 'powder of projection,' that used to obtain gold being of a red colour, while the powder for producing silver was white.

The processes for making the Stone recorded by some of the alchemists who claimed to have discovered it are usually so vague and wrapped up in such mysterious and symbolic language as to be unintelligible, but Rabbi Simon ben Canara, a

ALCHEMY AND ALCHEMISTS

Jewish alchemist of the sixteenth century, has left an account of his process for making the Philosopher's Stone which well illustrates the type of symbolism employed by writers of that period.

1

The mine of our Mercury, our Saltpetre, not of the vulgar, our Ponticity
Or purging quality and our Vitriol, not that of the vulgar,
Our actuating and our Sal Armoniack, not of the vulgar;

2

Of the germ of Mercury or living water or volatile salt,
Parts of the fixe, one part. The homogeneity of Mercury,
The living actuated, our Mercury, our living and Pontick water,
The Mercury of the wise, the homogeneous Mercury.

3

Let 2, or the most 3 parts of our Mercury liquefy,
One part of silver or gold subtiliated and they will become one body,
Spongious and inseparable which is called our silver and gold and not of the vulgar. The Silver by a Mercury of 5 eagles, the Gold by a Mercury of 7. This is the sophick calcination of Gold.

4

Irrovate or water our gold coming from our calcination with 2 parts of mercury in proportion to the gold vulgar. It will become a black body not spongious. The gold coming from the blackness at the change of colour openeth its germe and shows its vegetation.
This is the sophick Putrifaction and vegetative germination.

5

Our gold changing its green colour, becomes the white sulphur, which is a body animated with an incorruptable soul. This is our white gold without imposition of hands and opening of the vessel, melts or dissolves again with various vapours like the peacock's tayle by continuing of the heat. Then is our gold made volatile and this is a volatilization the wet way.

THE PHILOSOPHER'S STONE

6

Continuing the heat, our gold being made volatile, the wet way becomes dry and becomes a lucid body. The dawning of the day, and this is the volatilization the dry way. This dawning disappearing little by little. Our gold riseth in very fine red grains or powder which is our red sulphur so much wished by all the Sophi, but not the end of their labours. The volatilization the dry way.

7

Our red sulphur is imbibed with 5 parts of new mercury in proportion to the said sulphur's weight, and this by seven divers imbibitions and with continuation of the heat. By the rotation of the wheel of Nature, in a month's time is made putrifaction and at foresaid Regimens, at the end of the month, the King riseth Omnipotent which is our perfect Stone. The Medicine of the 3rd order, capable in its projection to transmute metals.

To God Eternal Praise. AMEN.

The following lines on the Philosopher's Stone are attributed to Basil Valentine, who will be referred to later:

> A Stone is found which is esteeméd vile,
> From which is drawn a fire volatile,
> Whereof our noble Stone its selfe is made,
> Composed of white and red that ne'er will fade.
>
> It's called a Stone and yet is no Stone;
> And in that Stone Dame Nature works alone,
> The Fountain that from thence did sometimes flow,
> His fixéd Father drownéd hath also.
>
> His life and body are both devoured,
> Until at last his Soul to him restored;
> And his volatile Mother is made one,
> And alike with him in his own kingdome.
>
> Himself also virtue and power hath gained,
> And far greater strength than before attained.
> In old age also doth the son excell
> His own mother who is made volatile

ALCHEMY AND ALCHEMISTS

By Vulcan's art, but first it's thus indeed
The Father from the Spirit must proceed ;
Body, Soul, Spirit are in two contained,
The total art may well from them be gained.

It comes from one and is one only thing,
The volatile and fixt, together bring
It is two and three, and yet only one,
If this you do not conceive you get none.

Adam in a Balneo resideth
Where Venus like himselfe abideth,
Which was prepared at the old Dragod's cost,
Where be his greatest strength and power lost.

It's nothing else, saith one Philosophus,
But a MERCURIUS DUPLICATUS.
I will say no more, its name I have shown,
Seek for it there, and spare no cost or pains,
The end will crown the work with health and gains.

The Elixir of Life is described by some as a solution of the Stone in spirit of wine and by others as a solution of gold, but the Elixir and the Stone were apparently often confused. By some they were considered to be one and the same body, but by others they are referred to as distinct substances. Some alchemists apparently believed that there were three different kinds of Stones—animal, vegetable, and mineral.

Artephius believed that human life could be prolonged for a thousand years by taking the Elixir, and others declared that it gave not only perpetual youth, but also an increase of knowledge and wisdom. Regular, daily doses of one grain dissolved in a sufficient quantity of white wine were to be taken in a silver vessel after midnight. To preserve health it was to be taken at the beginning of spring and autumn. "By this means," says Zacgarias, "one may enjoy perfect health until the end of the days assigned to him." Isaac of Holland recommended that the dose should be taken once a month; life would thus be prolonged "until the supreme end fixed by the King of Heaven."

For prolonging life and rejuvenating the aged, Paracelsus

THE PHILOSOPHER'S STONE

recommended his Primum ens Melissæ, which, he says, should be prepared by dissolving pure potassium carbonate in water

APPARATUS FOR DISTILLING AQUA VITÆ
From a woodcut
Brunschwig, 1507

and macerating in the solution the fresh leaves of the melissa plant. On this mixture spirit of wine was to be poured several times to absorb the colouring matter, after which the liquid was to be distilled and then evaporated down to the thickness of a syrup.

The 'Alcahest,' his celebrated medicine for all ills, was made with freshly prepared caustic lime and spirit of wine, which mixture was to be distilled ten times. The residue left in the retort was to be mixed with pure potassium carbonate and dried. It was then to be placed in a dish and ignited, and the residue formed his famous remedy.

CHAPTER X
FAMOUS ALCHEMISTS OF MEDIEVAL TIMES

MUCH of the story of alchemy may be gleaned from the lives of those who were engaged in its study and from the works they left behind them. We know that from the thirteenth to the fifteenth century their ranks were recruited from men of various classes, for Albertus Magnus, writing about the middle of the thirteenth century, says: "I find abbots, superiors, canons, physicians, and many unskilled folk who prosecuted this art." The majority appear to have been men who belonged to various religious orders, theologians, philosophers, and thinkers, who were drawn to the art not with the sole aim of making gold. These were the mystics, many of whom were enthusiastic and honest workers who sought knowledge by means of experiments and serious study, with the object, no doubt, of solving the mysteries that surrounded the prolongation of life and discovering the agent by means of which transmutation could be carried out.

The mystic alchemists made the spiritual motive their chief object and the making of gold secondary; some declared that they "thereby hoped to make gold so common that it would cease to have value for mankind." They believed that the true transformation or the transmutation of one thing into another could be effected only by spiritual means acting on the spirit of the thing, "because the transmutation consisted essentially in raising the substance to the highest perfection whereof it was capable." The result of the spiritual action might become apparent in the material form of the substance.

Then there were those who may be termed the pseudo-alchemists—the fraudulent rogues, unskilled and ignorant, who sought to trade on the credulity of others. These men were the charlatans who used the phrases of alchemy to deceive their dupes. The mystery by which the art was purposely surrounded and the mixture of magic and astrology combined to make the practice of the pseudo-alchemist a successful one, and those who consulted him were easily deceived by his pretended learning.

Some of the mystic alchemists, like Albertus Magnus (Albertus Groot), although they were not professed practitioners of the art were believers in it, and regarded it as a branch of science and philosophy. Albertus Groot was born at Lauingen, on the Danube, in 1193. He became a member of the Order of Dominicans in 1222, and afterward taught philosophy and theology in Hildesheim and Paris. In 1260 he was made Bishop of Regensburg, but after three years resigned the see and retired to a monastery at Cologne, where he died in 1280.

He wrote many works on alchemy, but some attributed to him are said to be spurious. He is stated to have been the first to use the word 'alkali' with reference to caustic potash, and he mentions lead oxide, iron sulphate, and cream of tartar in his writings. In a treatise entitled *Libellus de Alchemia* attributed to him he records his untiring search for the Philosopher's Stone and the Elixir of Life, which he apparently pursued in a real spirit of research. In commenting on those who had been lured to alchemy he remarks how many of them were reduced to poverty in their search, and gives the following rules and conditions that the alchemist should observe in his pursuit:

First. He should be discreet and silent, revealing to no one the result of his operations.

Second. He should reside in a private house in an isolated situation.

Third. He should choose his days and hours for labour with discretion.

Fourth. He should have patience, diligence, and perseverance.

ALCHEMISTS OF MEDIEVAL TIMES

Fifth. He should perform according to fixed rules.
Sixth. He should use only vessels of glass or glazed earthenware.
Seventh. He should be sufficiently rich to bear the expenses of his art.
Eighth. He should avoid having anything to do with princes and noblemen.

Considerable obscurity surrounds the life of Raymond Lully, whose name became famous in the thirteenth century. Many of the works attributed to him are now declared to have been written by another person of the same name. Elias Ashmole attributes to him the poem called *Hermes' Bird*, a very curious piece of alchemical lore.

The original Lully is said to have been descended from a noble Spanish family and was born in Majorca about 1235. He first devoted himself to the study of science, but when about the age of thirty he became a monk of the Order of the Minorites and eventually a missionary.

According to tradition, he took up the study of alchemy from the desire to cure a girl who was suffering from cancer. Afterward he wandered through Europe to acquire a further knowledge of the science.

There is a story told by Elias Ashmole that John Cremer, Abbot of Westminster, met him in Italy and persuaded him to come to London, that he worked in Westminster Abbey, where, a long time afterward, the cell which he occupied was discovered, and that in it was found a quantity of gold-dust. The Abbot of Westminster, who was also an alchemist and had been searching for the Philosopher's Stone for thirty years, told King Edward III about Lully.

Another story states that when Lully arrived in London lodgings in the Tower were assigned to him and that there he made gold. Constantinus, writing in 1515, declares that he actually saw "the golden pieces that were coined from the metal, which at that time were still named in England the nobles of Raymond or Rose nobles."

ALCHEMY AND ALCHEMISTS

As no record can be found of any Abbot of Westminster called John Cremer, both of these stories are probably fictitious.

Lully is said to have written a treatise on the Emerald Tablet, and in some of the works attributed to him methods of making nitric acid are described, together with its action on certain metals. He is also said to have made alcohol by distillation and to have known how to dehydrate it by the aid of potassium carbonate, which he obtained by calcining cream of tartar. He also recorded methods of making various tinctures and of obtaining essential oils from plants. In a fourteenth-century manuscript attributed to him, and now in the library of the Escorial, a list of the various names and synonyms of the substances used in his time are recorded.

In his later years he again devoted himself to missionary work and left Rome, where he had been living, to preach the Gospel in Africa. He was seventy-seven years of age when he landed at Bugia or Bona, in Algeria, to begin his mission, but, according to the story, he so irritated the Mohammedans when he landed by cursing their Prophet that they stoned him and left him to die on the seashore. His religious activities are still recorded in the island of his birth, and he left behind him a following of believers, the Lullists, who after a time spread over Europe.

Another alchemist of the mystical type was Thomas Aquinas, a man of deep religious principles. He entered the Order of the Dominicans early in life, and later studied under Albertus Magnus at Cologne and Paris. Of his work little is known, but from his writings it may be judged that he was averse from fraud and trickery. Like other alchemists of his time, he knew that there were substances that could make metals white, and that there were others, like sulphur and arsenic, that could give them a golden hue. Thus he says: "Add to copper some white sublimated arsenic and you will see the copper turn white. If you then add some pure silver you transform all the copper into veritable silver." He evidently believed that the base metals could be changed in nature by being deeply coloured. Brass he

THE STONING OF RAYMOND LULLY
From a woodcut (1515)

regarded as metal on the way to becoming gold, and copper whitened by silver as a metal being changed to silver.

That there were those in his time who were ready to use this knowledge for fraudulent purposes is evident from his observation that " to sell gold and silver made by the alchemist, if it has not the nature of true gold and silver, is fraudulent. If, however, it is true gold or silver the transaction is lawful."

Arnaldus de Villa Nova, who became celebrated in the thirteenth century, was contemporary with Raymond Lully. He is said by some to have been born about 1235 at Villeneuve-Loubet, near Avignon, but by others he is stated to have been a native of Spain. He was educated by the Dominicans and first took up the study of medicine, later going to Aix and thence to Paris, where Roger Bacon and Albertus Magnus were teaching philosophy. Here he remained for ten years, taking the degree of Master of Arts and acquiring a great reputation as a physician. To further his studies he went to Montpellier, which was then the centre of medical teaching of the Arabian school, and after completing his studies he travelled to Barcelona, whither his fame had preceded him, and was appointed chief physician at the Court of Aragon. It is said that when travelling in Italy he met Raymond Lully and helped to lay the foundations of Lully's reputation as an alchemist. Returning to Montpellier in 1289, he travelled on to Paris, where he became involved in religious controversies which ended in his being thrown into prison and the burning of his books. Through the intervention of the Archbishop of Paris he recovered his liberty, and in 1301 resolved to leave France and settle in Italy. Three years later we find him at the Papal Court of Benedict XI, after the death of whom he lived for some years in Sicily as physician to King Ferdinand, the brother of James, King of Aragon, who had always remained his friend. Summoned to the bedside of Pope Clement V at Avignon, he remained at the palace for some time until he was sent on a mission to Naples, where, it is said, he again met Lully and was able for a time to continue his work in alchemy.

In 1311 he returned to Paris, and when passing through Avignon on his journey he was offered the post of chief physician to Pope Clement V, who retained a great regard for him. He declined to accept the position and continued his journey.

ALCHEMISTS OF MEDIEVAL TIMES

Unfortunately he again got involved in religious troubles in Paris, and in fear of the Inquisition he left hurriedly for Sicily. During the voyage to Palermo the vessel in which he sailed was wrecked on the coast of Africa, but he managed to reach his destination in safety. While in Palermo he wrote a work on the School of Salerno which added to his fame. In 1313 he was again summoned to Avignon to attend Pope Clement and started off on the journey, but when near Genoa he was taken ill. He died on the ship and was buried at Genoa in 1313.

Arnaldus was a remarkable man in many ways. A friend and confidant of kings and popes, he was celebrated for his learning throughout most of the countries of Europe. As an alchemist he believed in the doctrine of transmutation and was a seeker for the Philosopher's Stone. He introduced the idea of the spirit of vital principle, or what was afterward called the quintessence, which became so notable a feature in the speculation of the mystics at a later period. He was one of the first to use alcohol for making preparations of drugs, and employed it in making tinctures and elixirs. He considered a solution of gold to be "a most perfect medicine," and as such it was regarded by some as the true Elixir of Life, a belief which persisted for centuries afterward.

It was not men of comparative poverty only who took up the study of alchemy, for we find some who had wealth and affluence among those who devoted their lives to the pursuit. Of the latter type was Bernard of Treves, who, from the age of fourteen until he was eighty-five, is said to have toiled incessantly in his laboratory in the quest for the Stone.

Born at Treves or Padua in 1406, the son of a wealthy man who left him a large fortune, Bernard became fascinated by alchemy, and when still a youth began to study the works of the Arabian alchemists with assiduity. When he was twenty he formed a friendship with a friar of the Order of St Francis who had similar tastes, and together they became persuaded that in highly rectified spirit of wine they would find the universal solvent that would aid them in transmutation. They worked for three years and spent three hundred crowns before they discovered that they were mistaken. They next turned their attention to investigating alum and copperas, but were again unsuccessful. Attracted by their labours, adepts, to whom Bernard gave a share of his wealth, came from all parts to aid them.

After losing his friend the friar Bernard was joined by a magistrate of Treves who had formed a belief that the sea was the mother of gold, and that sea-water was capable of changing lead and iron into the precious metal. Impressed by this idea, Bernard resolved to investigate it and transported his laboratory to the shores of the Baltic, where he worked on salt for over a year, but, in spite of his experiments in melting, subliming, and crystallizing it, he found he was no nearer his goal.

When he reached the age of fifty he determined to travel and seek the experiences of other alchemists. He first went to France, where he remained for five years, when, hearing that one Master Henry, confessor to the Emperor Frederick III, had discovered the Stone, he set out for Germany. Accompanied by five other adepts he met on the way, he duly arrived in Vienna, where he was entertained by the alchemists of that city. He gained an interview with Master Henry, who frankly told him that, although he had been searching for the Stone all his life, he had not found it, but was prepared to go on searching until he died.

Master Henry proved to be a man after Bernard's own heart, and they vowed each other eternal friendship. At their instance the alchemists of Vienna were got together, and at a meeting it was resolved, as all were equally interested in the quest, that each man present should contribute a certain sum toward raising forty-two marks of gold. Master Henry confidently told them that if they did so he would in five days increase the gold fivefold by means of his furnace. Bernard, ever generous, contributed ten marks as his share, while Master Henry and the others gave one or two each.

ALCHEMISTS OF MEDIEVAL TIMES

On the day appointed for the great experiment all assembled in a laboratory, and the gold marks collected were placed in a crucible together with a quantity of salt, copperas, *aqua fortis*, mercury, lead, eggshells, and excrement. The company gathered round and watched with intense interest, fully expecting that when taken from the furnace the crucible would be found to contain molten gold. But alas, after persevering with the experiment for three weeks, they found that the gold in the crucible amounted to the value of sixteen marks, instead of the forty-two which had been put in.

Although Bernard made no gold in Vienna he parted with a great deal, and in discouragement gave up his search for two months. Later, hearing that a famous alchemist in Rome had discovered the secret, he set out for that city. He was again disappointed, and journeyed on to Messina and thence took ship to Cyprus, Greece, and Constantinople. For eight years he wandered through Palestine and Egypt and eventually reached Persia. The journey, he tells us, cost him 13,000 crowns, and, becoming short of money on his return, he had to sell a portion of his estates. It is said that he next visited England and stayed four years, but eventually he returned to Treves at the age of sixty-two, an impoverished man.

His relatives, who regarded him as mad, refused to have anything to do with him, but, still confident of ultimate success in his quest, he retired to the island of Rhodes. Here he met a monk who also was an enthusiast in alchemy, but neither had the money to carry on experiments. After negotiating for a year Bernard at last found a merchant in Rhodes who agreed to advance him 8000 florins on the security of the last remaining portion of the alchemist's estates. He recommenced his labours with renewed vigour, and ate, slept, and lived in his laboratory. He still dreamed of his work being crowned with success when he had almost come to his last coin, and at the age of eighty starvation stared him in the face. There is a legend that he at last succeeded in discovering the secret of transmutation when in his eighty-second year, and that he lived for three years longer, enjoying the wealth it afforded him until he died in 1490. According to his own account, written in his last years, he had made a more important discovery than that of the Stone in at length finding the secret of contentment.

Bernard's life affords an interesting illustration of the zealous labours of some of the workers in alchemy who were lured by the quest for the transmuting agent which always eluded them. They travelled from country to country at a time when such journeys were difficult and arduous, and doubtless suffered great hardships and privations in pursuit of a chimera which ever baffled them but ever led them on.

Little is known of the life of Trithemius, another mystic philosopher and theologian, who afterward became famous as the instructor of Paracelsus. He is said to have been born at the village of Tritheim, near Treves, in 1462. His father, John Heidenberg, was a poor vine-grower in the district and died before his son reached the age of seven. Until he was fifteen the son continued to work in the vineyards, but devoted his nights to the study of Greek and Latin, with the ambition of going to a university to take a course in philosophy. This he did, remaining at the university until he was twenty, when a great desire came upon him to see his mother, and one day he set out on foot for his native village. As he approached Spannheim, late on a winter's evening, a heavy snowstorm came on and so blocked the road that he could proceed no farther. At length he stumbled on a monastery, where he found refuge for the night, but the snow continued to fall for seven days and rendered the roads impassable. The hospitable monks would not hear of his departure and persuaded him to stay on with them. He became so impressed with their manner of life that he resolved to enter their order and renounce the world. He was gladly received as a brother, and grew so beloved that in the course of two years the brethren elected him as their abbot.

The monastery at that time was falling into ruin, but Tri-

AN ALCHEMIST
From a drawing in a manuscript of the fifteenth century

SYMBOLIC FIGURE

It represents "The Blessed Stone—the Sun and Sulphur with hys Mercurie, bothe Bodi and Soul to God and Man."

From an alchemical manuscript of the fifteenth century

ALCHEMISTS OF MEDIEVAL TIMES

themius took affairs in hand and, by good management, not only restored the building, but reorganized the community. He set the brethren to work to copy manuscripts of the great philosophers and ancient writers until they possessed a fine library.

He remained Abbot of Spannheim for twenty-one years, spending much of his time in the study of alchemy and the occult sciences. He is reputed to have practised both magic and sorcery, and there is a story that he raised from the grave the spirit of Mary of Burgundy at the instance of her husband, the Emperor Maximilian. Eventually Trithemius became Abbot of St James at Würzburg, and remained there until his death in 1516.

The monasteries at this period were the nurseries of the sciences, and the fame of Trithemius spread throughout the country, many students journeying to Spannheim to learn from him the secrets of the arts. Among them was Paracelsus, who is said to have remained at the monastery for some time and gathered many secrets from the Abbot.

The story of Nicholas Flamel, who is supposed to have lived in Paris between 1330 and the end of that century, is one of romantic interest. He is said by some to have been a fictitious personage, and there are certainly doubts whether he ever existed; on the other hand, it is probable that the story of his life may have been woven round some alchemist who flourished at a later date.

What is known of him is contained in an account of his life which is said to have been discovered after his death. He tells us that he earned his living as a scrivener in Paris and that he lived in the Rue Notaire, near the church of Saint-Jacques-la-Boucherie, in the year 1399. After the death of his parents Nicholas continued to carry on his work of writing and engrossing inventories and other documents, and in the course of his avocation he one day met with a very large and ancient gilded book, which he bought for two florins. The leaves of the book appeared to him like the rinds of young trees, and the brass covers bore a curious kind of lettering which he took to be Greek or some ancient language. The book contained thrice seven leaves, so numbered at the top of each folio, and on every seventh leaf were painted images and figures instead of writing. On the first leaf of the first section was depicted a virgin who was being swallowed by serpents, on that of the second section was a cross upon which a serpent was crucified, while on that of the last section a wilderness, watered by many fair fountains out of which issued a number of writhing serpents, was represented. The first leaf of the book contained the following inscribed in great characters of gold: "Abraham the Jew, Prince, Levite, Astrologer, and Philosopher, unto the Jewish nation scattered through France by the wrath of God; wishing health in the name of the Lord of Israel." Thereafter followed great execrations and maledictions against anyone who should glance within unless he were a priest or a scribe.

The first leaves contained consolations to the Jewish race, and the last of the written leaves dealt with the transmutation of metals, which the author said he set down in order to help his captive nation in paying tribute to the Roman emperors. No mention, however, was made of the prime agent necessary for transmutation, but he indicated that he had figured and emblazoned it with great care on the fourth and fifth leaves.

Nicholas then describes some of the illuminated figures on those leaves, which included a representation of the god Mercury, a fair flower on the top of a mountain, a rose-bush in flower in a garden, and a king carrying a falchion causing his soldiers to destroy a multitude of little children. Perceiving that the latter depicted the slaughter of the innocents by Herod, and believing that he had learned the main part of the art from the book, Nicholas resolved that he would place in the cemetery of the Holy Innocents these symbolic representations of the secret science.

Being unable to interpret the other emblematical pictures, Nicholas was at length induced to show the book to one Anselm,

ALCHEMISTS OF MEDIEVAL TIMES

a licentiate in medicine and a deep student of the art. He professed to solve them, but, his interpretations being more subtle than true, Nicholas states that after experimenting for twenty-one years he gave up all hope of understanding the figures.

At last he resolved to set out on a pilgrimage to the monastery of St James, in Spain, in the hope of meeting some Jewish priest who might have a key to the enigma. The journey was in vain, but when returning he met at Leon a merchant of Boulogne who introduced him to a Master Canches, a doctor of great learning who was a Christianized Jew. At the sight of the figures which Nicholas had copied from the book he was ravished with wonder and joy. He began at once to decipher them, and agreed to accompany Nicholas back to Paris to see the originals, but on reaching Orleans he fell sick and died.

Nicholas arrived safely back in Paris, and was welcomed with joy by his wife, Perrenelle. He told her that he had at least gained some knowledge from the Jewish doctor, especially of the "first matter." He then set to work, and after experimenting for three years he at length succeeded in making the "prime agent," which was revealed to him by a "strong odour." On the first occasion he used it (January 17, 1392) he transmuted half a pound of mercury into pure silver in the presence of his wife. On April 25 he made a projection of the Red Stone on the same amount of mercury and changed it into the same quantity of pure gold: he carried out this operation three times.

Nicholas thus soon became very wealthy and resolved to devote his riches for the good of mankind. Before the end of 1413 he and his wife had founded fourteen hospitals, built three chapels, provided seven churches with gifts, and restored cemeteries in both Paris and Boulogne.

In concluding his story Nicholas adjures those who wish to attain the inestimable possession of the "Golden Fleece" not to purchase earthly possessions with the proceeds, but to use them in helping their brethren and in relieving the poor and sick, widows and orphans.

ALCHEMY AND ALCHEMISTS

The reputed tombstone of Nicholas Flamel, which bears the date 1418 and is said to have been found in the old church of Saint-Jacques-la-Boucherie, is now preserved in the Cluny Museum in Paris.

There is a tradition, related by a chronicler, that Nicholas Flamel gave some of his Red Stone to a nephew of his wife named Perrier, and that from him it passed to a Dr Perrier. On his death it was found among the effects of his grandson, named Dubois. Dubois exhibited it to Louis XIII, and with it transmuted some base metal into gold in the presence of the King. He was then asked to make some more of the Red Stone, which he promised to do. However, he failed to produce it and, according to the story, was hanged in consequence.

ALCHEMISTS AT WORK IN A LABORATORY
From a woodcut
Brunschwig, 1507

CHAPTER XI

ENGLISH ALCHEMISTS OF THE MIDDLE AGES

THE first Englishman associated with alchemy appears to have been Robert of Chester, who, in the twelfth century, completed a translation of *The Book of the Composition of Alchemie*, which he made from the Arabic while in Spain. The original work is attributed to Marianus. It is not, however, until the following century that the foundations of natural science in England were laid by Roger Bacon. This scientist was born at Ilchester, in Somersetshire, in 1214, and after studying at Oxford became a friar. He devoted much of his time to the study of alchemy, and it is only within the present century that the extent of the work he accomplished has been recognized.

He defined alchemy as the formation of things from elements, and appears to have believed that potable gold was the secret of the true Elixir of Life. He is said to have been the discoverer of gunpowder, which, Hine says, he describes in an anagram which alludes to a powder producing a thunderous noise and a bright flash and made by mixing seven parts of saltpetre, five of young ashwood, and five of sulphur. In his work entitled *De Secretis Artis Naturæ*, written in 1249, he gives a process for refining saltpetre, and in other treatises deals with the metals gold, silver, lead, tin, copper, and iron. He describes their properties and various processes for subliming, distilling, and calcining, and treats of the management of furnaces and the regulation of heat. He spent a large amount of money on instruments and apparatus in order to carry out his experiments, and his works show that he was a man of great intelligence and an earnest seeker after scientific truth.

ALCHEMY AND ALCHEMISTS

Reference to two alchemists of the fourteenth century is made in the Plea Rolls of that period, one being John de Walden, who in 1350 was imprisoned in the Tower because he failed in his alchemical experiments "on 5000 crowns of gold and 20 pounds of silver, which he had received from the King's treasure to work thereon by the art of alchemy for the benefit of the King." The other is "William de Brumley, chaplain, lately dwelling with the Prior of Harmandsworth." He was arrested by order of the King's Council, with four counterfeit pieces of gold upon him. He expressly acknowledged that he had by the art of alchemy made these pieces from gold and silver and other medicines—to wit, "sal armoniak," vitriol, and "golermonik" (probably persulphuret of tin). The process had occupied him five weeks, and he had taken the pieces to Gautron, the keeper of the King's money at the Tower, and offered to sell them if they appeared to him of any value. William had before sold to Gautron a piece of this sort of metal for 18s., but of what weight it was he did not know. He said that he made the metal according to the teaching of William Shuchirch, Canon of the King's Chapel at Windsor. Two separate juries, one of six laymen and the other of three experts, valued the four pieces offered by William Brumley at 35s., but they declared them not to be pure gold.

Of the English alchemists of the fifteenth century George Ripley, who was a canon of Bridlington, is perhaps best known. A contemporary writer tells us that "he was a man of a quick and curious wit who spent almost his whole life in searching on the occult and abstruse causes and effects of natural things." On giving up his monastic life he set out to travel Europe, and journeyed through France, Germany, and Italy, where he visited the Court of Innocent VIII. From Italy he sailed to the Isle of Rhodes, residing there for some months with the Knights of the Order of St John of Jerusalem, to whom, it is said, he gave £100,000 annually toward maintaining the war that they were then carrying on against the Turks. On returning to England he became a member of the Carmelite Order and renewed his alchemical studies. His chief work, *The Compound of Alchymie conteining Twelve Gates*, dedicated to King Edward IV, was written in 1475, and is in verse. In it he describes the twelve processes necessary for the achievement of the *magnum opus*, which he likens to the twelve gates of a castle which the philosopher must enter. These he names Calcination, Solution, Separation, Conjunction, Putrefaction, Congelation, Cibation, Sublimation, Fermentation, Exaltation, Multiplication, and Projection. He states that alchemists purposely employ mystery in order "to discourage the fools, for although we write primarily for the edification of the disciples of the art, we also write for the mystification of those owls and bats which can neither bear the splendour of the sun nor the light of the moon."

"Concerning the first gate," Calcination, he writes:

> Calcination is the purgacyon of our Stone
> Reatauryng also of hys naturall heate.

Of the second gate, Solution, he says:

> Here a secret to thee I wyll dysclose
> Whych ys the ground of our secrets all.
> . . . . .
> Take hede therefore in Errour that thou not fall,
> The more thyne Erth and the lesse thy Water be
> The rather and better Solucyon shall thou see.

Of the third gate, Separation:

> Separacyon doth ech parte from other devyde,
> The subtill to the Groce, fro the thyck the thyn.

Of the fourth gate, Conjunction:

> Of two conjunctions philosophers doe mentyon make
> Groce when the Body with Mercury ys reincendat.

> Putrefaction, the fifth gate, may thus defyned be,
> After Phylosophers sayings it ys of Bodyes the fleyng.
> And in our Compound a dyvysyon of thyngs thre
> The Kylling Bodyes into corrupcyon forth ledyng.

In similar vague language he continues to his twelfth gate, Projection, of which he says:

> If my Tyncture be sure and not varyable,
> By a lyttyl of thy medcyn thus shall thou prove
> Wyth mettall or wyth Mercury as Pyche yt wyll cleve
> And Tynct in Projeccyon all Fyers to abyde
> And sone yt wyll enter and spred hym full wyde.

He completes his verse with the admonition,

> Now thou hast conqueryd the Twelve Gates
> And all the Castell thou holdyst at wyll
> Kepe thy secrets in store unto thy selve
> And the commandements of God looke thou fulfill.
> In fyer conteinue thy glas styll
> And Multiply thy medcyns ay more and more
> For wyse men done sey store ys no sore.

Ripley certainly succeeds in mystifying his readers, and whatever truth there may be in his verse is enveloped in symbolism. Some of his works are in the form of long scrolls, drawn and coloured, full of symbolic and emblematic figures of men and animals, which are supposed to represent certain alchemical processes. The texts are so vague they appear to be almost meaningless, and are said to have been revealed to him in a vision. In *The Compound of Alchymie* he thus describes some of his experiments:

> Many amalgam did I make,
> Wenyng to fix these to grett avayle
> And thereto sulphur dyd I take
> Tartar, eggs whyts and the oyl of the snayle.
> But ever of my purpose dyd I fayle,
> For what for the more and what for the lesse,
> Evermore something wanting there was.

Later Ripley obtained an Indulgence from Pope Innocent VII exempting him from "Claustrall observance," and he left the Carmelite Order, and is said to have become an anchorite until his death in 1490.

Thomas Norton, who was born in Bristol, belonged to an old family long associated with that city. His father held the office of sheriff in 1401 and was elected mayor in 1413, while he also represented Bristol in Parliament. The house where the Nortons lived, a fine old half-timbered building, now called St Peter's Hospital, is still standing.

Thomas became an earnest student of alchemy when a young man, for he tells us that he had arrived at the knowledge of preparing the Elixir of Gold when he was twenty-eight. It is said that he became a pupil of George Ripley, and that he achieved more than local fame as an alchemist is shown by the references made to his work by many writers of a later date.

According to an account of his life which he gives in one of his works, he once rode over a hundred miles to meet his master and stayed with him forty days, in the course of which he learned all the secrets of alchemy. On returning to Bristol he renewed his studies and succeeded in making the great Elixir of Gold, but the secret was stolen from him by a dishonest servant. Undaunted by this loss, he continued his labours and at length discovered the much-sought Elixir of Life; but again misfortune came upon him, for this secret also was stolen from him by a merchant's wife, who is supposed to have been the spouse of William Canynges, a Bristol merchant who rebuilt the famous church of St Mary Redcliff.

Norton's chief work, called the *Ordinall of Alkimie*, was printed in the *Theatrum Chemicum Britannicum* (1652) of Elias Ashmole, who states that he made the transcript from a very fine manuscript which he had compared with fourteen other copies. He found the author's name from the first word of the Proheme and the initial letters of the six following chapters.

> Thus we may collect the author's name and place of residence, for these letters together with the first line of the seventh chapter speak thus:
> Tomais Norton of Briseto
> A parfet master ye maie him trowe.

Such like fancies

were the results of the wisdome and humility of the aunciet philosophers who (when they intended not absolute concealment of Persons, Names, Misteries, etc.) were wont to hide them by Transpositions, Acrostiques and the lyke.

In the Proheme Norton shows how the craze to make gold at that time had seized on all classes. He writes:

> Common workmen will not be out-lafte
> For as well as Lords they love this noble craft;
> As gouldsmithes whom we shulde lest repreve
> For sights in their Craft moveth them to believe,
> But wonder it is that Wevers deale with such warks
> Free-masons and Tanners with poore Parish Clerks;
> Tailors and Glasiers will not thereof cease,
> And eke sely Tinkers will put them in the prease.

In a beautiful manuscript copy of Norton's *Ordinall of Alkimie* in the British Museum, probably written about 1477, there are some finely executed miniatures showing alchemists at work in their laboratories. In one an alchemist is depicted seated at a table on which stands a pair of scales enclosed in what appears to be a glass case or box, almost a counterpart of the chemical balance used to-day.

The work itself is interesting as showing how mysticism and science were intertwined by the alchemists of the time, who regarded the art as a kind of divine revelation, or what Norton calls the "subtle science of holi alchyme."

He begins by asking:

> Shall all men teache?
> What manner of people maie this science reache,
> And which is the true science of alchemis
> Of ould fathers called blessed and holi?

He warns his readers that the secrets must not be committed to writing:

> It must need be taught from mouth to mouth
> And also he shall (be he never so loath)
> Regard it with sacred and most dreadful oath.
> So blood and blood maie have no parte
> But only vertue wynneth this HOLI ART.

MINIATURE OF THE FIFTEENTH CENTURY DEPICTING
AN ALCHEMIST IN HIS LABORATORY

On the table at which he is seated is a chemical balance, and in the foreground two assistants are superintending operations with an aludel, or sublimatory.

Norton's *Ordinall of Alkimie* (B.M. MS.)

MINIATURE OF THE FIFTEENTH CENTURY REPRESENTING AN
ALCHEMIST'S LABORATORY IN WHICH TWO ASSISTANTS ARE
ENGAGED IN VARIOUS OPERATIONS

Norton's *Ordinall of Alkimie* (B.M. MS.)

ENGLISH ALCHEMISTS

Like the earlier alchemists, Norton believed that the true secrets of the art were only revealed to certain holy men, for he says, "Almightie God from great doctors hath this science forbod and grant is to few men of his merret; such as be faithful true and holi."

He thus alludes to the ancient legend of the origin of alchemy which was doubtless commonly believed at the time:

> This science beareth her name by a King called Alchumis,
> A glorius Prince of most noble minde,
> His noble virtues helpt him this Arte to find.
> He searched Nature; he was a noble clerk,
> He sought and found this Arte.
> King Hermes also did the same,
> Which was a clerk of excellent fame.
> In quadripartite manner of astrologie,
> Of physick and this Arte of alchemie,
> And also Magic naturall.

Concerning Raymond Lully Norton says:

> Such as truly make gold and silver fine
> Where of example for testimony,
> In a cittie of Catalony
> Which Raymond Lully, knight as men suppose,
> May in seven images the truth disclose.

He next proceeds to tell the story of one Thomas Dalton, a monastic alchemist who had succeeded in making the Philosopher's Stone, for

> Of the red medicine he hath great store.
> I trow never Englishman had more.

He tells us that Dalton was taken out of an abbey in Gloucestershire against his wish and brought before King Edward. Thomas Herbert, body-squire to the King, was sent to bring Dalton to Court, and Delius, his confidential squire, reported to the King that Dalton had made a thousand pounds' worth of good gold.

The King asked Dalton to perform the process again, and questioned him from whom he had the secret. He replied, from a

ALCHEMY AND ALCHEMISTS

canon of Lichfield. The King then gave Dalton money to go and find the canon where he would. Dalton set forth on his quest, but was waylaid and kidnapped by Herbert, who lay in wait for him and brought him to Stepney. They stole all the money the King had given him and eventually took him to Gloucester Castle, where they kept him a prisoner for four years and then brought him out to die.

Norton continues:

> This was his payne as I you tell
> By men who had no dread of hell.
> Herbert died soon after in his bedd
> And Delius at Tewkesbury lost his head.

The third chapter of the book begins:

> For the love of one
> Shall briefly disclose the matters of our store
> Which the Arabians doe Elixir call
> Wherefor it is here understand you shall.

In it he mentions one Tonsill, evidently an apothecary, who had

> spent all his lustie daies
> In fals Recipts and such lewde assayes,
> Of herbes, gummes, of rootes and of grass
> Many kindes by him assayed was,
> As crowfoot, celandine, and Mizerion,
> Vervayne, Lunara and Martagon,
> In Antimony, Arsenick, Honey, wax and Wyne.

The fourth chapter:

> Teacheth the greate worke
> A foule labour, not fitt kindlie for a clerke
> In which is found greate travaile
> With many perills and many a faile.

Here he mentions a "subtill balance," probably the one enclosed in a case of glass depicted in the miniature facing p. 96. He refers to "Albertus Magnus, the blackfryer," and says that neither he "nor Bacon Fryer minor, his co-master, had not our red stone in consideration." He concludes his book by

> Praying all men which this book shall find
> With devout prayers to have my soul in mind.

ENGLISH ALCHEMISTS

Norton was the author of other works entitled *De Transmutatione Metallorum* and *De Lapide Philosophico*. From the writings of his great-grandson, Samuel Norton, who was also an alchemist, we learn that Thomas was a member of the Privy Council of King Edward IV.

Of John Dastin, another English alchemist, little is known. Bale, the historian, speaks of him as the "prime alchemist of his age and the only Master thereof in England." He is said to have been the author of several treatises on the art. One of these is supposed to have been revealed to him in a wonderful dream in a book written in letters of gold. In this treatise, wrapped in mysterious and symbolic language, the story is told of "a mighty rich King" and of "the wedding of the Sun and Moon and the feast that followed, at which all the planets attended."

William Bird, who was Prior of Bath and who helped to build the Abbey Church, was another of the monastic alchemists. He was the instructor of Thomas Charnock, who declares that he actually discovered the Elixir and hid it in a wall, but alas, "when ten days after he went to fetch it out he found nothing but the stople of a cloute. He lost his sight and so was deprived of attempting to make the Elixir again, whereupon he lived obscurely and grewe very poore."

Thomas Charnock, who is described by Ashmole as "a student in the most worthy scyence of astronomy and philosophy," was the author of a book called *A Breviary of Naturall Philosophy* which was printed in 1557. That he also practised alchemy is evidenced from the following lines in which he alludes to a monk of Salisbury who was famous as an alchemist:

> For all our wrettings are verye darke
> Despyse all Bookes and them defye
> Wherein is nothing but Recipe and Accipe.
> Few learned men within this Realme
> Can tell the aright what I do mean;
> I could finde never man but one,
> Which cowlde teache me the secrets of our Stone,
> And that was a Pryste in the Close of Salusburie.
> God rest his Soll in heaven full myrie.

ALCHEMY AND ALCHEMISTS

The beliefs of other English alchemists of this period, such as John of Preston, who wrote *The Mirror of Elements*, William

AN ALCHEMIST BLOWING THE FIRE OF A FURNACE
From a woodcut
Schrick, 1500

Bloomfield, and others of whom we know nothing but their names, may be summarized in these anonymous lines (1477):

> A wonderful science and gifte of Almightie
> Which was never founde by labour of man.
> But it by teaching or revelation began.

ENGLISH ALCHEMISTS

Elias Ashmole gives an interesting description of a curious stained-glass window, placed behind the pulpit in St Margaret's Church at Westminster in the fifteenth or sixteenth century, which symbolized alchemy but, being mistaken later for a Popish story, was broken to pieces. He says:

> The window is divided into three parts. In one was a man holding a boy in his hand and a woman with a girl in hers all standing upright, naked postures, upon a greene foliate earth.
> Both had fetters on their feet and seemed chained to the ground, which fetters were presented as falling from their legs.
> Over their heads were placed the Sun and the Moon painted of a sad darke colour.
> On the left side of the window was a beautiful young man clad in a garment of various colours bearing a yellow cross upon his shoulders, his body encircled with a Bright Glory which sent forth beams of divers colours. He stood upon an earth intimating *oculus piscium*.
> In the lower and middle part of the window was a faire large red rose which issued rays upward and in the middle an exceeding bright yellow glory. Above was the figure of a man rising with beams of light spread about his head. He had a garment of a reddish colour deepened with red and heightened with yellow. In his left hand he had a white stone which he held towards the persons arising in that part of the window, and in his right hand he held a red stone towards him whose garments were of various colours.
> Over the figures was inscribed, "OMNES GENTES ADEPTI PLAUDITE QUIA DOMINUS FRATER VESTER."

At the bottom of the window were several coats of arms, "but," says Ashmole, "after very diligent search among the Records of English coats of arms these could not be found."

It is evident from the date of his writing that the window was still in St Margaret's Church in the first half of the seventeenth century. In some of the early symbolical drawings Adam and Eve represented gold and silver or the sun and the moon, and the white and red stones held by the figure "rising with beams of light" were obviously intended to represent the Philosopher's Stone.

CHAPTER XII

ALCHEMY IN THE TIME OF JOHN GOWER AND GEOFFREY CHAUCER

FROM the fourteenth to the sixteenth century the alchemist played a part in the life of the English people, as is shown by the frequency with which he is introduced into the literature and plays of the period.

Both John Gower and Geoffrey Chaucer allude to the deceptions that were being practised in their time by the pseudo-alchemists and to the pretentiousness of their claims. From the knowledge of the science they show in their works it is probable that they both had had some practical experience in the art. Ashmole, indeed, declares that Chaucer "knew the mystery" and was "a judicious philosopher and is ranked among Hermetic Philosophers, and his master in this science was John Gower." He says that the first acquaintance between them began at the Inner Temple, where "Sir John studied the laws and whither Chaucer came to follow the like course of studies upon his return out of France. They became great friends, and soon perceived the similitude of their manners and quickly joined in friendship and labours."

Gower appears to have believed in the doctrine of the 'three stones,' vegetable, animal, and mineral. He thus alludes to the doctrines of alchemy in the *Confessio Amantis*:

> These olde philosophres wise
> By wey of kinde in sundry wise
> Three stones made through clergy.
> The firste if I shall specify,
> Was called *vegetabilis*,
> Of which the propre vertue is
> To mannes hele for to serve
> As for to kepe and to preserve

ALCHEMY IN THE TIME OF GOWER

> The body from sikenesses alle,
> Till deth of kinde upon him falle.
> The stone seconde I thee behote
> Is *lapis animalis* hote,
> The whose vertue is propre and couth
> For ear and eye and nose and mouth,
> Wherof a man may here and se
> And smelle and taste in his degre.
>
> . . . .
>
> The thirdde stone in speciall
> By name is callèd *minerall*,
> Which the metalles of every mine
> Attemper till that they ben fine,
> And purify them by such a wey
> That all the vice goth awey
> Of rust, of stinke and of hardnesse.
> And whan they ben of such clennesse,
> This minerall, so as I finde,
> Transformeth all the firste kinde
> And maketh hem able to conceive
> Through his vertue and to receive
> Both in substaunce and in figure
> Of golde and silver the nature.

Elias Ashmole describes the pseudo-alchemists as

quacking Mountebanks, nibbling sciolists, and ignorant juglers. Let philosophers say what they can and wise men give never so good counsell no warning will serve, they must be couzened, nay they have a greedy appetite thereunto; but it has been ever so, so strong and powerfull a misleader is covetousness.

Norton describes these cheats, Ripley dissects them to the bone and scourgeth them naked to the view of all,

> For covetous men that findeth never
> Though thy seke it once and ever.

In *The Canon's Yeoman's Tale*, written about 1390, Chaucer throws further light on the methods of the pseudo-alchemists, who were evidently common at the time. It will be remembered that the pilgrims, while ambling along on their way to the city, were overtaken by a man clad in black and very shabby. He was a canon of the Church, and was followed

ALCHEMY AND ALCHEMISTS

by his yeoman, who attached himself to the cavalcade and began to tell the travellers what a wonderful man his master was and how he could make all the gold he required. The Canon tries to stop his gossip and then rides off, while the Yeoman relates his tale of the disappointments he had met with in practising the art, declares that alchemists are fools, and tells of

> the Philosophres Stone,
> Elixer called, we seek faste everyone,
> . . . .
> For all oure craft, whan we have al y-do,
> With al oure sleighte, he will not come us to.

From the subsequent observations of the Canon's Yeoman, it would appear as if some sudden resentment had determined Chaucer to interrupt the regular course of his work in order to insert a satire against the alchemists.

He thus proceeds to tell of the work in which the Yeoman engaged and how he blew the fire till his heart grew faint:

> What should I tellen each proportiôn
> Of thinges whiche that we work upon ;
> As on five or six ounces, may well be,
> Of silver, or some other quantity ;
> And busy me to telle you the names
> Of orpiment,[1] burnt bones, iron squames,[2]
> That into powder grounden be full small !
> And in an earthen pot how put is all,
> And salt put in, and also peppér,
> Before these powders that I speak of here,
> And well y-covered with a lamp of glass
> And muchel other thing which that there was,
> And of the pot and glasses enlutyng [3]
> That of the air mighte passe out no thing,
> And of the easy fire, and smart also
> Which that was made, and of the care, and woe
> That we had in our matters subliming
> And in amalgaming and calcining,
> Of quick-silver, called mercury crude ;
> For all our sleightes we can not conclude.

---

[1] Trisulphide of arsenic.   [2] Scales.   [3] Sealing with lute.

ALCHEMY IN THE TIME OF CHAUCER

> Our orpiment and sublimed mercurie,
> Our grounden litarge [1] eke on the porfurie,
> Of each of these of ounces a certáin,
> Nought helpeth us, our labour is in vain.

The formula here given, consisting of arsenic, bone-ash, and iron scales, was for an amalgam well known to alchemists of the time. He continues:

> There is also full many another thing
> That is unto our craft appertaining,
> Though I in order them not rehearse can
> Bycause that I am an untaught man;
> Yet will I tell them as they come to mind,
> Though I ne can not set them in their kind—
> As bool Armenian,[2] verdigris, borax,
> And sundry vessels made of earth and glass;
> Our urinals and our decensories,[3]
> Vials, crucibles, and sublimatories,
> Distilling flasks and alembics eke,
> And other such, that are not worth a leek—
> Not needeth it for to rehearse them all
> Waters rubifying and bull's gall,
> Arsenic, sal ammoniac, and brimstone,
> And herbes could I tell each many one,
> As agrimony, valerian, lunary [4]
> And other such, if I should choose to tary.

He then proceeds to mention other materials such as

> Salt of Tartar, alkali, and sal-preparat,
> And combusted matter and coagulate,
> Rat's bane [5] and our matter enbibing,
> And eke of our matter embodying,
> And of our silver citrinacion,[6]
> Our cementing and fermentation,
> Our ingottes, testes, and many more.

He concludes with an allusion to the doctrine of the 'four spirits' and the 'seven bodies,' which has been called the alchemist's creed.

---

[1] White lead.   [2] Armenian bole or clay.
[3] Vessels employed for distilling *per descensum*.
[4] Moonwort.   [5] Arsenic.   [6] Turning lemon-colour.

ALCHEMY AND ALCHEMISTS

    I will you tell as was me taught also
The four spirits and the bodies seven,
By order, as oft I heard my lord them neven.[1]
    The first spirit quick-silver called is,
The second orpiment, the third I wis
Sal-ammoniac, and the fourth brimstone.
The bodies seven eke, lo, them hear anon :
    Sol gold is, and Luna silver we threpe,[2]
Mars iron, Mercury quick-silver we clepe,
Saturnus lead, and Jupiter is tin,
And Venus copper, by my father's kin !
    This cursed craft whoso will exercise
He shall no good have that shall him suffice ;
For all the good he spendeth thereabout
He lose shall—thereof I have no doubt.
Whoso that listeth out of his folly
Let him come forth and learn to multiply !
    .   .   .   .   .   .
But everything that shineth like to gold
Is not gold, as that I have heard it told,
Nor every apple fair unto the eye
Is not good, whatsoe'er men shout and cry.

Goethe, in *Faust*, thus pictures the alchemist:

    Who in his dusky workshop bending
    With proved adepts in company
    Made, from his recipes unending,
    Opposing substances agree.

Later he gives the following description of a process in the symbolic language of the art:

    There was a Lion red, a wooer daring,
    Within the Lily's tepid bath espoused,
    And both, tormented then by flame unsparing,
    By turns in either bridal chamber housed.
    If then appeared, with colours splendid,
    The Young Queen in her crystal shell,
    This was the medicine—the patient's woes soon
        ended,
    And none demanded—who got well?

Goethe is said to have drawn this description partly from Paracelsus and partly from Welling's *Opus Mago Cabbalisticum*.

    [1] Name.   [2] Call.

ALCHEMY IN THE TIME OF CHAUCER

The "Lion red" is cinnabar, called "a wooer daring" on account of the rapidity with which it united intimately with other bodies. The "Lily" is a preparation of antimony which was known as *Lilium Paracelsi*. In alchemical language red was the masculine and white the feminine colour. The glass alembic, which consisted of the two chambers (body and head), in which the substances were placed, was first put in a "tepid bath" and gradually heated, "tormented then by flame unsparing," the two bodies being drawn from the "bridal chamber" in the lower part to another, meaning that their wedded fumes were forced by the heat into the upper chamber, or head. If then the "Young Queen," or the sublimated substance, appeared with brilliant colours in the upper chamber, the proper result was obtained, and this signified the true "medicine."

John Lyly, in his comedy *Gallathea*, written in 1592, thus satirizes the alchemist. Raffe, a simple-minded fellow who is shipwrecked on a strange shore, meets with Peter, an alchemist's boy, who tells him of the wonders of alchemy and the greatness of his master. "A little more than a man," he declares, "and a haires bredth lesse than a god. Hee can make thy cap gold, and by multiplication of one grote three old angels." Peter assures Raffe that alchemy is a very secret science, for "none almost can understand the language of it, and it has as many termes impossible to be uttered as the arte to be compassed." Much to Raffe's delight the alchemist consents to take him into his service, in which he hopes soon to get rich, but he quickly tires of the hard work in the laboratory and abandons his master no richer than he was before.

CHAPTER XIII

THE ALCHEMIST'S LABORATORY AND HIS APPARATUS

THE laboratory or workshop of the alchemist has been made familiar to us by the Dutch and other painters of the seventeenth century. David Teniers, Witt, Van der Veldt, Jan Steen, and other masters found in these picturesque and dark interiors, with their varied array of curiously shaped apparatus, subjects of never-failing interest.

They delighted to portray the bearded and venerable alchemist, in cap and gown, intent on some operation in his quest for the Stone, or seated in a chair peering at some manuscript or ancient tome that might perhaps give him the key to the mystery. From the black and smoke-begrimed beams and rafters of the roof hangs the crocodile, alligator, or strange-shaped fishes, and from a corner glares a great owl, symbolic of wisdom. On the shelves that line the walls stand quaint jars, curiously shaped flasks and bottles, or the skulls and bones of beasts, birds, and fishes. On one side of the adept stands the great still, and on the other his aludel or sublimatory, while on a bench or table close by are his globe and hour-glass. The floor is littered with jars, pots, dishes, crucibles, mortars and pestles, funnels, tongs, and other implements of his art. Away in the background the assistant is often seen working the great bellows, feeding a furnace, or luting some apparatus and making ready for the next operation. The twisted condenser, like a monstrous snake, and the big globular recipients or long-necked matrasses are tinged with gleams of light from a burning brazier, while alembics glow like fireflies in a dim corner.

The adept sometimes began his operations with prayer, as we

THE ALCHEMIST'S LABORATORY

learn from the *Mappæ Clavicula,* in which the recipe for "Making of Gold" begins, "Prayer you are to recite during the operation or the fusion that follows, in order that the gold may be formed." The idea of prayer doubtless arose from the invocations to deities that came down from the times of the Egyptians, which in the Middle Ages came to be merged into Christian practices.

But the alchemist's surroundings could not have been always the picturesque interiors which artists loved to represent, for, according to descriptions of laboratories that have been left by writers of the sixteenth and seventeenth centuries, they were usually arranged with workmanlike precision.

PLAN AND ELEVATION OF A LABORATORY
Libavius, 1606

The various types of apparatus that have come down to us from the days of the alchemists form an interesting study. Some of them led to important discoveries in times gone by and are now almost forgotten, while others still survive in our laboratories and have changed but little in shape throughout the centuries.

The earliest known representation of a piece of apparatus used in alchemy appears in the Greek manuscript of a treatise attributed to Synesius and said to have been written in the fourth century. It is the drawing of a crude still, the body of which is formed by a large cucurbit, a bulbous-shaped vessel round at the bottom with a short, wide neck, surmounted by a caput, or head, from which a tube runs into a receiver.

Several of the early manuscripts on alchemy contain drawings of the apparatus described in the texts, and many of these types were probably designed about the end of the tenth or the begin-

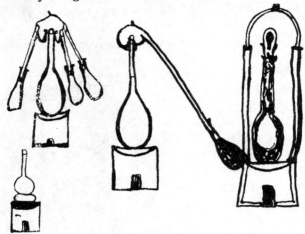

TRIBICUS, STILL, AND OTHER APPARATUS DESCRIBED BY SYNESIUS
IN THE FOURTH CENTURY
From a drawing in a Greek manuscript

ning of the eleventh century. Others may date from an earlier period, as they correspond to descriptions of apparatus recorded

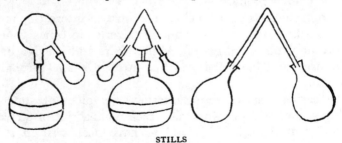

STILLS
From a thirteenth-century manuscript

by Zosimus and Olympiodorus, and among them we have stills, flasks for digestion, and a Balneum Mariæ, which was a vessel for holding hot water in which a flask or retort containing the

AN ALCHEMIST IN HIS LABORATORY
David Teniers the Younger (1610–90)
Photo Bruckmann, Munich

THE ALCHEMIST
Thomas Wyck
Photo Bruckmann, Munich

THE ALCHEMIST'S LABORATORY

substance to be digested was kept at a gentle heat for the necessary period.

The still appears to have been originally built up from a cucurbit; from this evolved the alembic, the name of which was given by the Arabs to the complete still from the Greek *ambix*,

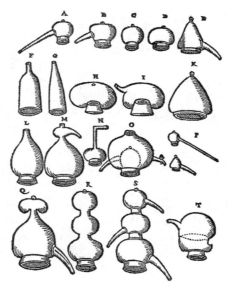

STILL-HEADS AND ALEMBICS
Libavius, 1606

a cup. At first built up from two pieces with a removable head, at a later period the alembic, when it was of glass, was sometimes made in one. Alembics were also made of clay and earthenware to withstand greater heat. The recipient or receiver, which was fitted over the open end of the tube of the alembic in order to receive the distillate, was a globular vessel which varied in size and was usually of glass.

Another early piece of apparatus was the retort, which appears to have been originally adapted from the matrass with the tube curved downward. It had various forms according to the angle at which the tube was bent, either high or low. Retorts were made of clay, earthenware, or glass. The matrass was a flask with a long neck that was employed for many purposes.

The pelican, another piece of apparatus frequently used, was a vessel with a tubular neck and provided with two beaks, one

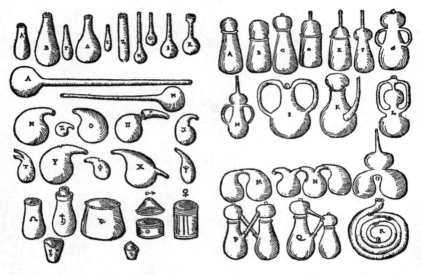

RETORTS AND MATRASSES
Libavius, 1606

PELICANS, CUCURBITS, AND TWINS
Libavius, 1606

opposite to the other, in order to conduct the vapour to the lower part of the vessel, so that distillation could be carried on constantly.

The aludel, or sublimatory, dates from about the thirteenth century and, as its secondary name implies, was used for sublimation. It consisted of a gourd-shaped pot of earthenware or clay, so fashioned that a series could be built one upon the other to a height of five or six feet. The bottom pot was set on a stove, which supplied the heat, and as the vapour arose it condensed on the inside of the upper pots, from which it was afterward removed by scraping.

THE ALCHEMIST'S LABORATORY

Another important piece of apparatus was the athanor, which usually formed a prominent feature of the laboratory. It was a type of furnace made of clay, and stood about five feet high. It was constructed so that it could be divided into several parts, the bottom one consisting of a small fireplace which supplied the

APPARATUS FOR DIGESTION OR SUBLIMATION
From a Syriac manuscript of the thirteenth century in the British Museum

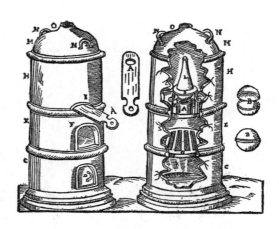

ATHANORS
"The athanor was the reverberatory oven of the philosophers. The fire did not touch the base, and the required heat was suitably and uniformly imparted."—RULAND.
Libavius, 1606

necessary heat. It was employed chiefly for the purpose of digestion, but was also used for evaporating certain liquids.

The serpent, used in distillation, was usually made of metal, and consisted of a spiral or zigzag tube which acted as a condenser and connected the cucurbit with the caput. The balloon was a spherical-shaped vessel, usually made of glass, with two or three short spouts or beaks, and was employed to collect the distillate, the third spout being set in a recipient to receive it. The cloche, the forerunner of the bell-jar, was used for subliming, while the crucifix, which was built up of three cucurbits round a central glass vessel, was employed for the same purpose. The

ALCHEMY AND ALCHEMISTS

philosopher's egg was the name usually given to a flask with a round bulb and a short neck by which it could be attached to another flask. The evaporating dish was a circular-shaped

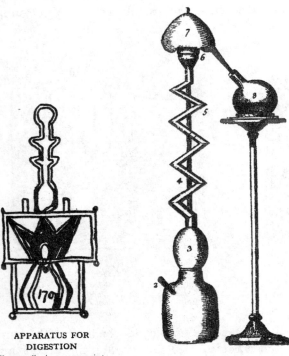

APPARATUS FOR DIGESTION
From a Syriac manuscript of the thirteenth century in the British Museum

SERPENT CONDENSER
Eighteenth century

shallow vessel which was placed over the Balneum Mariæ or a sand-bath and in which liquids were placed for evaporation.

The earliest form of condenser appears to have been the serpent or zigzag, which was cooled by air. The water-cooled coil that passed into a vat does not appear to have been used until the fifteenth or sixteenth century. The type of condenser associated with the name of Liebig was known at a much earlier

THE ALCHEMIST'S LABORATORY

period, and, judging from an ivory carving owned by the writer, must have been used at least two centuries ago. This object is a beautiful ivory mortar of the seventeenth century; it was formerly in the Goldsmid Collection, and the carving in high relief represents the interior of a laboratory in which

FURNACES OF VARIOUS TYPES
Libavius, 1606

several boys are manipulating various pieces of apparatus. One boy holds a staff entwined with a serpent, emblematic of medicine, two are using a pestle and mortar, two more are attending to a pump, while another holds a spouted flask. A furnace and a still are shown, and to a large retort is attached a condenser, cylindrical in shape, with a funnelled tube near the centre into which water can be poured; a bent tube on the other side is probably intended for an outlet. At the end of the condenser is a vertical tube through which the distillate passes into the receiver.

There were various types of furnaces, such as the reverberatory used for the distillation of mineral substances, the distillation furnace for distilling *per descensum*, and the furnace for fusing

APPARATUS FOR DISTILLATION
Libavius, 1606

metals. Such operations as the evaporation of extracts and saline solutions were generally conducted on the alchemist's hearth. Baths, both wet and dry, were employed for the digestion of substances at various degrees of heat. For very slow evaporation sand was used instead of water, while cinders or iron filings on a slow furnace were sometimes employed for the same purpose.

Caput mortuum was the term given to the residue that was left in the bulb of a retort after an operation, and the word

THE ALCHEMIST'S LABORATORY

'cohobation,' often met with in works on alchemy, was applied to the repetition of distillation, as when the distillate was poured on the material from which it had already been distilled and then redistilled.

Important to the alchemist when erecting his apparatus were the substances, or mixture of substances, called lutes, by means

A RETORT AND RECIPIENT STILL FOR MAKING SPIRIT OF WINE
From *The Art of Distillation* (1667)

of which he put together, fixed, and sealed the various parts. Soft clay seems to have been the basis of many of them, but the formulæ for lutes in treatises on alchemy are very numerous. "To have good lutes in a laboratory," says a writer of the sixteenth century,

> is a thing very necessary for all, for which I furnish you herewith the following lessons about it, what lutes and clays receive ye fires and what do not. First, a luting for still or receiver, loam serves well enough but not for ye fires, for in a strong fire it melts and is not fit therefore for use. Loam mixed to a compost with horse-dung till soft enough can be used for glass and other retorts

and crucibles, but Stourbridge clay, mixed with powdered glass and water to the consistency of a soft paste, makes the finest lute of all.

Glass-house sand and loam mixed together are said to make another good lute, while clay tobacco-pipes reduced to powder, calcined, and then mixed with clay, form a lute which does not

DISTILLING APPARATUS
Woodcut
From *Ryff*, 1567

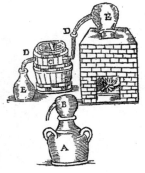

A HOT STILL
From *The Art of Distillation*
(1667)

spoil in the fire and is therefore very suitable for all glass. A common lute for glass vessels was composed of flour and whiting, mixed into a paste with water. This was applied very thinly to the joints in the apparatus.

In the following interesting account, which was found attached to a manuscript written in 1560, the prices of apparatus supplied to an alchemist at that period are shown.

2 bottel bodies	pro	3 solidis	6	denarius
3 Alembici	,,	4 ,,	2	,,
6 quart Retort	,,	4 ,,	6	,,
1 bottel Retort	,,	1 ,,		
6 pint Retort	,,	3 ,,		
1 bottel head } 1 quart head }	,,	2 ,,	6	,,
5 parting glasses	,,	2 ,,		
1 recepter	,,	8 ,,		
3 Phialæ	,,	1 ,,	6	,,

THE ALCHEMIST'S LABORATORY

2 Cucurbitæ cum 2 Alembicis cæcis pro 3 solidis
2 Receptacula pro retort   ,, 3 ,,
2 alia vitæ   ,,   9 denarius

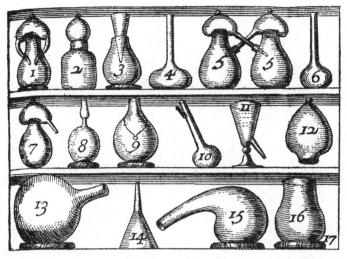

APPARATUS EMPLOYED IN A LABORATORY IN THE LATE SEVENTEENTH CENTURY

1, Pelican; 2, vessel for mixing; 3, hell; 4, flat-bottomed matrass; 5, twins; 6, matrass; 7, alembic in one piece; 8, philosopher's egg; 9, egg within an egg; 10, small matrass; 11, separation glass; 12, blind alembic without a beak; 13, recipient; 14, glass funnel; 15, retort; 16, cucurbit; 17, disc or straw mat upon which vessels were placed.

It will be observed from the above illustration, which is taken from a French work printed at the end of the seventeenth century, that the shapes of some pieces of apparatus underwent changes in the course of time.

Although some types have since become obsolete, the matrass, now generally called a flask, the retort, the recipient or receiver, and the funnel are still employed by chemists in their laboratories to-day.

CHAPTER XIV
ALCHEMICAL SYMBOLS AND SECRET ALPHABETS

THE symbols employed by the alchemists to represent the various elements, metals, substances, processes, and operations used in their art, that developed to such a great extent in the Middle Ages, go back to an early period. Originally designed to ensure secrecy in recording the alchemists' formulæ and to prevent the uninitiated from acquiring their knowledge, they also served as a kind of shorthand for working out their recipes and notes just as the symbols used by chemists do to-day. Secret alphabets, ciphers, and emblematic drawings representing various processes also served as a medium of understanding between the adepts of different nations and became universally employed in Europe.

Joseph Scaliger, the historian, says, "The symbols and signs are of great antiquity and are often found engraved on ancient tombs and runic stones." Some show evidence of having been derived from ancient Egyptian sources, while others, like the symbols used for the metals, had their origin in the signs employed to represent the planets, the heavenly bodies, and the Zodiac, and are of an even earlier period.

The disks representing the sun and the moon and the symbol of Venus are depicted on a Babylonian boundary-stone which dates from 2500 B.C. The association of certain deities with the planets had its origin with the Sumer-Akkadians, probably before 4000 B.C., and the astral-mythological cult elaborated by the Babylonians later penetrated the Orient and thence passed to Egypt, Greece, and Rome.

The supposed connexion between the heavenly bodies and

SYMBOLS AND SECRET ALPHABETS

the metals, as already shown, is attributed to early astrological influences. To the Babylonians the seeds of the metals were in the earth, and the influence of the Sun-god produced gold, the Moon-god produced silver, and so with the other metals known to them. This idea became one of the basic doctrines of alchemy and persisted until the seventeenth century. The signs of the Zodiac, which the early astrologers considered almost of equal importance with those of the planets, were undoubtedly the origin of some of the alchemical symbols as instanced in the sign for Libra, ♎, which in alchemy was used to signify sublimation, and the symbol for copper, ♀, is but the Egyptian *crux ansata*. Other alchemical symbols derived from the Zodiacal signs are those for saltpetre, borax, marcasite, cinnabar, and realgar. The sign for Leo became the symbol for alum, sublimate, lead, and vinegar; Sagittarius for antimony, iron, glass, and salt; Pisces for sal ammoniac and amalgam; Aries for zinc, fire, and wine; Gemini for potash, sand, vitriol, and tutty; Aquarius for water, sublimated arsenic, and white lead; Scorpio for a human skull; and Taurus for alum, verdigris, and copper.

The astrological nomination of the metals and the mystic relation with the number seven existed in the Far East, and the same idea of the seven heavens, each with its own gate of a different metal, is evidenced in Babylonia and Persia.

Of the alchemical symbols derived from ancient Egyptian sources we have that of the sun used to represent gold, the moon for silver, the wavy line for water, and the symbol for mercury, which has a resemblance to the hieroglyph for Thoth, whom the Greeks called Hermes and the Romans Mercury.

The majority of the symbols used by alchemists afterward were of Greek origin, and are found in manuscripts on alchemy which date from the eighth to the eleventh century. The Greek

ALCHEMY AND ALCHEMISTS

EARLY GREEK ALCHEMICAL SYMBOLS

1, Gold (the sun); 2, electrum; 3, silver (the moon); 4, copper; 5, iron; 6, lead; 7, tin; 8, mercury; 9, sulphur; 10, arsenic; 11, oil; 12, salt; 13, cadmium; 14, earth; 15, vinegar; 16, magnesia; 17, arsenic; 18, chalk; 19, iron; 20, vermilion; 21, natron; 22, cinnabar; 23, cadmium; 24, green; 25, blue; 26, red; 27, vitriol; 28, sea-water.
From manuscripts of the eighth to tenth centuries

manuscript now preserved at St Mark's, in Venice, contains a large number of symbols of which the following are examples:

GOLD   SILVER   COPPER   IRON   LEAD   TIN   SULPHUR   MERCURY

The Arabian alchemists rarely used symbols, and it is not until about the fifteenth century, when the European alchemists were searching for the Philosopher's Stone, that they multiplied and came into general use. As time went on many of these symbols were altered or added to by alchemists of different nationalities, and so their number was greatly increased.

It is interesting to compare some of the symbols used at various periods, and the following selection covers examples found in manuscripts from the eleventh to the seventeenth century.

The symbols usually employed to represent the elements were:

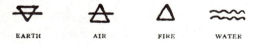

EARTH   AIR   FIRE   WATER

The earliest-known symbol for gold, the circle, appears to have been used from an unknown period to signify perfection and

STILL FOR MAKING OIL OF VITRIOL
From a Spanish manuscript on distillation, 1588

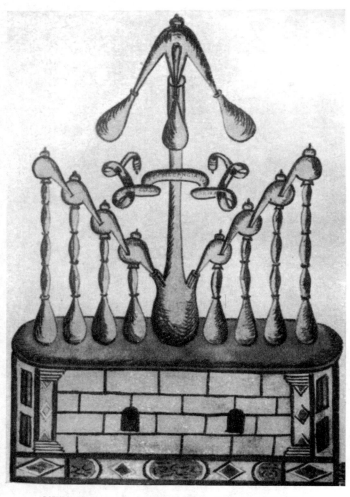

MULTIPLE STILL FOR MAKING THE ELIXIR DI
MARA VIGLEOSA VIRTU

From a Spanish manuscript on distillation, 1588

SYMBOLS AND SECRET ALPHABETS

simplicity, and is represented on runic and stone monuments of great antiquity.

Besides the simple circle, there are at least sixty-three other known symbols for gold that were used between the twelfth and the eighteenth centuries, of which the following examples are arranged chronologically:

O ☉ ⚈ ℮ ⌯ ♃

The earliest symbol used for silver, which was governed by the moon, was a simple crescent; this later went through twenty-six variations:

) (☽ ⚭ ☉ ♍

The symbol for mercury was that employed to represent the planet, and it went through fifty-seven variations:

☿ ♀ ⚴ ☽♃ ♈

The symbol for copper, which was adapted from the sign for Venus, consisted of a cross beneath a circle, denoting that the body of the metal was like gold but joined with some corrosive substance the symbol for which was a cross or part of a cross. There were forty-six variations of the symbol:

♀ ⚥ ⊶ ⚤ ⚥ ⊞ ♃

The symbol for iron was derived from the sign for the planet Mars, and is said to have originated from the idea of a shield and a spear symbolizing the God of War. There were thirty-two variations of this symbol:

♂ ⊶ ♁ ⚬ ⇒ ⚳

The symbol for tin is said to have originated in the sign of

ALCHEMY AND ALCHEMISTS

invocation to the planet Jupiter, which was associated with the metal. This symbol had thirty-four variations:

Lead, which was associated with the planet Saturn, was represented by the sign for that planet, and had forty-one variations. Its corrosive quality was believed to be very powerful; therefore the cross on the figure was placed in a superior position.

The symbols for antimony are also numerous, and are at least fifty-six in number:

The symbols for some of the principal substances used by the alchemists were as follows:

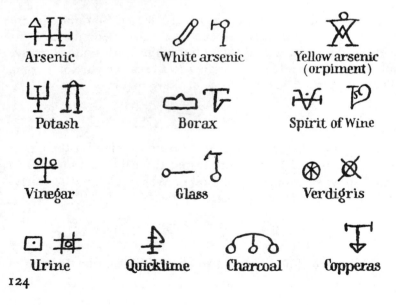

Arsenic	White arsenic	Yellow arsenic (orpiment)
Potash	Borax	Spirit of Wine
Vinegar	Glass	Verdigris
Urine	Quicklime	Charcoal	Copperas

SYMBOLS AND SECRET ALPHABETS

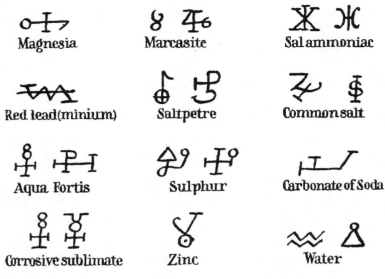

Magnesia · Marcasite · Sal ammoniac
Red lead (minium) · Saltpetre · Common salt
Aqua Fortis · Sulphur · Carbonate of Soda
Corrosive sublimate · Zinc · Water

Operations were represented by the following symbols:

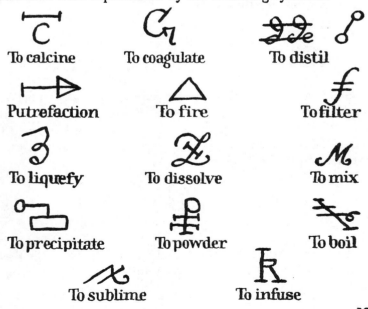

To calcine · To coagulate · To distil
Putrefaction · To fire · To filter
To liquefy · To dissolve · To mix
To precipitate · To powder · To boil
To sublime · To infuse

ALCHEMY AND ALCHEMISTS

The following symbols represent apparatus and appliances:

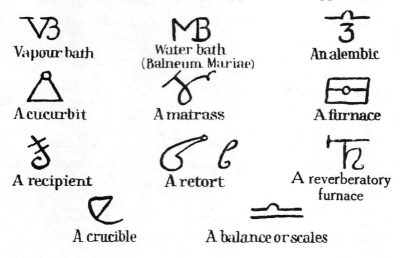

The symbols for time were:

Some alchemists invested the symbols with a special or hidden meaning. Thus, Glauber represented them in squares, and explained that the extent to which the symbol touched the four sides indicated how near it approached to perfection, as instanced in the following examples:

From the Middle Ages to the eighteenth century the literature on alchemy that arose was immense, and hundreds of

SYMBOLIC REPRESENTATION OF ONE OF THE "TWELVE KEYS OF THE GREAT STONE OF THE PHILOSOPHERS"

A MONASTIC LABORATORY
Both of these subjects are from engravings of the seventeenth century

SYMBOLIC FIGURE REPRESENTING THE GREEN LION DEVOURING THE SUN

SYMBOLIC FIGURE REPRESENTING COAGULATION OF THE FIRST STONE AND ITS SUBLIMATION

From *The Rosary of the Philosophers*, a manuscript of the seventeenth century

SYMBOLS AND SECRET ALPHABETS

manuscripts were written on the subject. The symbols and secret characters to guard the precious secrets and hide them from the vulgar were constantly being altered. They multiplied in every country, and thus render their processes and formulæ very difficult to decipher.

Operations and processes were sometimes represented by emblematic pictures and drawings, but even the symbolism in these frequently varied. In these pictorial representations the figure of a king clad in red was employed to signify gold, and that of a queen in white symbolized silver. A yellow lion was sometimes used to represent the yellow sulphides, a red lion for cinnabar, a green lion for salts of copper or iron, an eagle or a crow for black sulphides, a salamander for fire, a lion with wings and one without represented mercury and sulphur, and a man and a lion the earth. The green lion represented the mercury of the philosophers, and a wolf antimony. The figure of a dragon being killed by the sun and moon symbolized mercury being combined with gold and silver, while a green lion devouring the sun signified mercury dissolving gold.

The preparation of the Philosopher's Stone was sometimes

SYMBOLIC FIGURE REPRESENTING A PROCESS IN ALCHEMY
From a manuscript of the fifteenth century

represented by the figure of an infant, and a tree bearing the sun, moon, and five stars was used to symbolize 'universal matter.' Putrefaction was sometimes represented by a decomposing body, a skeleton or a crow; sublimation by birds flying upward; precipitation by birds descending; conjunction by a marriage scene; and perfection by the picture of a baby.

SYMBOLIC ALCHEMICAL FIGURE REPRESENTING A PROCESS; THE DRAGON HAS BEEN RESUSCITATED AFTER DEATH
Nazari, 1572

In describing their operations and processes in symbolical language such allegories as death, burial, and putrefaction were frequently used, and for the perfection of substances the wedding of male and female. Putrefaction was essential, as "no metallic seed can develop or multiply unless the said seed by itself alone and without the introduction of any foreign substance be reduced to a perfect putrefaction."

In the works of Basil Valentine the processes wherein the agent effects the perfecting of the less perfect things are divided into stages which he calls the gates. The first is calcination, or drying up; the second dissolution, likened to death and burial; the third conjunction, when the separated substances are combined; the fourth putrefaction, necessary for the germination of the seed; the fifth congelation, gathering together, or cohesion of the substances; the sixth sublimation, in which the body flies upward; the seventh fermentation, when the substance becomes soft and fluid; and the eighth exaltation, which is the perfection or essence. To these may be added cohobation, which meant repeated distilling. The distillate was

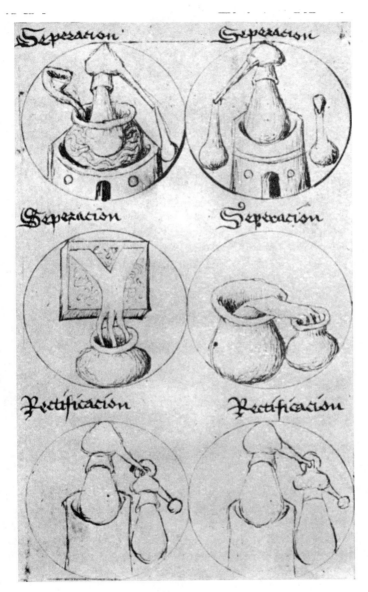

DRAWINGS OF APPARATUS FOR SEPARATION AND
RECTIFICATION
From an alchemical manuscript of the fifteenth century

SYMBOLIC REPRESENTATIONS OF OPERATIONS
1, Sublimation; 2, Solution; 3, Putrefaction; 4, Fermentation;
5, Separation; 6, Fixation
From drawings in a manuscript of the sixteenth century

SYMBOLS AND SECRET ALPHABETS

put back into the retort and redistilled in order to concentrate or increase the amount of virtue in it.

The alchemists firmly believed that the most effectual method of separating a complex substance into more simple ones was to subject it to the action of heat; therefore the majority of their processes consisted of distilling and redistilling, incinerating or subliming, in order to obtain what they called the essence or true spirit. Certain important operations were sometimes depicted emblematically in the form of flasks, and described as follows:

(1) *Division.* In this art is the separation of the parts of the compound; represented by the sun and moon over flames.

(2) *Coitus.* This is the natural act of two, suppose of body and spirit; represented by a lion partly disappearing into a sphere.

(3) *Calcination.* This is pulverization of a dry matter by fire.

(4) *Sublimation.* This is the elevation of dry matter by fire, when the extract is driven to the highest part of the vessel; represented by the moon in the earth and a bird above ascending into the sky.

(5) *Solution.* This is the reduction of dry matter in a liquid; represented by the sun and moon in darkness and a bird descending from the sky.

(6) *Generation.* In matter and form is only contained the generation of nature; represented by a bird descending into a coloured sea.

(7) *Putrefaction.* This is the corruption of the proper and natural heat in every moist thing; represented by a bird descending into darkness.

(8) *Fermentation.* This is the incorporation of the inanimate or that which giveth life, the restoration of taste, the inspiration of smell; represented by a bird descending into water in which is a black toad.

(9) *Conjunction.* This is the joining of two or more bodies; represented by a white bird descending into flames with air, water, and earth below.

(10) *Separation.* This is the division or separation of bodies; represented by a bird ascending into the air with fire and earth below.

(11) *Imbibition.* This is a soaking or maceration of a body in a liquid; represented by flames ascending into the sky or a dragon in flames with his tail in his mouth.

(12) *Fixation.* " In this art the body receives the tingent spirit and takes away its volatilnesse. Out of perfection of the Fixing the fire is gold by ye help of whom only is the stone."

"Secrecy," says Michael Maier, the alchemist and mystic, referring to alchemical symbols and emblems,

> is sometimes secured by using entirely dissimilar characters for one article or one operation. The arbitrary nature of these symbols has necessitated keys explaining them. One of these was discovered on 6th May, 1403, in the walls of the cloister connected with the church at Schwartzbach, by an adept in alchemy who hid them again in the cloister at Marienzell, Thuringia, but they were rediscovered on 10th December, 1489.

Novices were bound by a sacred oath to keep all operations secret, and every precaution was taken, by means of symbols, ciphers, and secret alphabets, to prevent experiments and processes from becoming common possessions.

Secret alphabets and ciphers were employed for recording processes which were regarded as of special importance. Three of these alphabets, taken from manuscripts of the fifteenth and sixteenth centuries, are here reproduced.

The symbols employed by the alchemists for the metals and their salts were passed on to the chemists in the eighteenth century, and some of them are still to be seen represented on the carboys in the windows of the old chemists' shops in various parts of the country.

At this period, when chemistry was beginning to develop into a science, new substances and compounds were discovered and additional characters had to be devised. Therefore many of the original symbols were modified.

Bergmann, who was professor of chemistry at Upsala, invented a system by which, using some of the old symbols and modifying others, he claimed to be able to represent almost every substance then known to chemists.

A B C D E F G H I K L M N

O P Q R S T U X Y

A B C D E F G H I K L M N

O P Q R S T U

A B C D E F G H I K L M N

O P Q R S T U V Y Z

SECRET ALPHABETS USED BY ALCHEMISTS IN THE FIFTEENTH AND SIXTEENTH CENTURIES

SOME ALCHEMICAL SYMBOLS USED DURING THE MEDIEVAL PERIOD

Key to the List of Alchemical Symbols

1. Air
2. Earth
3. Fire
4. Water
5. Antimony
6. Yellow sulphuret of arsenic
7. Red sulphuret of arsenic
8. Potash
9. Oil
10. Lead
11. Borax
12. Spirit of wine
13. Iron
14. Vinegar
15. Calamine
16. Spirit
17. Glass
18. Gold
19. Verdigris
20. Wine
21. Quicklime
22. Copper
23. Copperas
24. Aqua vitæ
25. Magnesia
26. Marcasite
27. Red lead
28. Powder
29. Mercury
30. Sal ammoniac
31. Saltpetre
32. Common salt
33. Aqua fortis
34. Sulphur
35. Silver
36. Stones
37. To sublime
38. Human skull
39. Tutty
40. Oil of vitriol
41. White vitriol
42. Tartar
43. Bismuth
44. Bismuth
45. Zinc
46. Tin
47. Cinnabar
48. To decoct
49. Balneum Mariæ
50. Cucurbit
51. To calcine
52. To coagulate
53. To distil
54. Alembic
55. To digest
56. To filter
57. Putrefaction
58. Matrass
59. Precipitate
60. To prepare
61. Quintessence
62. Reduction
63. Retort
64. Reverberatory furnace
65. To liquefy
66. To boil
67. An hour
68. To sublime
69. To infuse
70. A week
71. Soap
72. Sulphur of the philosophers
73. Crucible
74. Crystal
75. To amalgamate
76. A balance
77. Borax
78. Day
79. Night
80. Litharge
81. Copper ore
82. Charcoal
83. Wax
84. Alum

CHAPTER XV
SOME NOTABLE ALCHEMICAL MANUSCRIPTS

THE earliest treatises on alchemy in Latin date from about the eighth century, and two in particular, which are supposed to have been written in the ninth and tenth centuries, are of considerable interest.

The work entitled *Compositiones ad tingenda* contains a number of recipes which were apparently used in the arts and crafts of the time, which lends colour to the theory that in early times alchemy was largely influenced by the workers in metals. In this manuscript we find formulæ for dyeing skins, for gilding iron, for writing in letters of gold, and for soldering metals. It also contains recipes for colouring artificial stones used in making mosaics, directions for staining glass and the staining of wood, bone, and horn, together with a list of the ores, metals, earths, and metallic oxides employed by craftsmen in making jewellery, in enamelling, and in painting. Instructions are given for gilding glass, wood, metals, and fabrics. Mention is also made of gums, resins, and other vegetable substances used in the arts, of products derived from the sea, such as salt, coral, and molluscs yielding a purple dye. A formula for making bronze is given, and a recipe for writing in letters of gold is similar to one recorded in a papyrus at Leyden, which shows that in all probability many of these recipes date back to a much earlier period.

In the treatise called the *Mappæ Clavicula*, the earliest-known copy of which is said to date from the tenth century, several recipes are given for making 'gold,' and in a later copy of this

NOTABLE ALCHEMICAL MANUSCRIPTS

work, supposed to have been written in the twelfth century, the preparation of alcohol is thus described:

> On mixing a pure and very strong wine with a third of a part of salt, and heating it in vessels suitable for the purpose, an inflammable water is obtained which burns away without consuming the material (on which it is poured).

Lippmann believes that alcohol was first obtained by distillation in Southern Italy in the twelfth century.

In a manuscript of the thirteenth century gold is described as a "metallic body, citrine, ponderous, mute, fulgid, equally digested in the bowels of the earth and very long washed with mineral water."

The majority of alchemical manuscripts of the fourteenth and fifteenth centuries are written in cryptic or symbolic language; this recipe for making the "Panasea of Gold" is an example:

> Take ye winged virgin newly alighted from ye chariott of ye sunne beams. Draw her lyppes over her teeth y^t her appetite may be increased. Putt her to bed to a sonne of her own would be begotten by her. Draw ye curtaines y^t they may sweate.

Sometimes recipes are given in verse, so that the student should be able to fix them in his memory, as instanced in these lines:

> Take
> Mercury and silver fine
> Wch in ammalgam combine.
> Then wch sublimate and vinegre fine
> Y^t must wash, penetrate and finily joyne.

Many manuscripts on alchemy are illuminated in gold and colours, especially those that illustrate various processes. In some a king, the sun, representing gold, is to be found arrayed in an elaborate costume of crimson, while a queen, the moon, symbolizing silver, is clad in blue. Some very fine examples are still preserved in the Bibliothèque Nationale and the Library of the Arsenal in Paris, in the principal libraries in Italy, such as the Vatican, the Ambrosian, and the Laurentian, and also in the Manuscript Department of the British Museum.

There is probably no more beautiful specimen of the illuminator's art in existence than the manuscript entitled *Splendor Solis*, now in the British Museum, which is said to have been written about 1582. The twenty-two full-page miniatures which illustrate it, each of which has a mystic meaning, are said to have been painted by the famous master Lucas van Leiden, and are exquisite examples of his art.

Space will not permit detailed description of this script of seven parts, in which is set forth the hidden mystery of the philosophers, alchemy and astrology being intermixed. It states that

> all corporal things originate in and are maintained and exist of the earth according to time and influence of the stars and planets, as Sun, Moon, and the others. These, together with the four qualities of the elements which are without intermission, moving and working therein, thereby creating every growing and procreating thing in its individual form, sex, and substance, as first created at the Beginning by God, the Creator, consequently all metals originate in the earth of a special and peculiar matter produced by the four properties of the four elements which generate in their mixture the metallic force under the influence of their respective planets.

In this we have the gist of the work, and the illustrations are chiefly devoted to the sun, the father and the sulphur of the philosophers, and the moon, the mother of the Philosopher's Stone.

The work is supposed to have been written by Solomon Trimosin, an alchemist who is said to have instructed Paracelsus in the art and imparted to him the great secret. A probably fictitious account of his life and adventures is related in *Aurum Vellus*. We are told that, like others who pursued the art, Trimosin while quite a young man set forth in 1473 in quest of knowledge, and in his travels came in contact with one Flocker, an alchemist and miner, who was believed to possess the great secret. He is said to have used a process with common lead, adding to it a peculiar sulphur.

AN ALCHEMIST
From a miniature by Lucas van Leiden in *Splendor Solis*

GOLD AND SILVER REPRESENTED BY A KING AND QUEEN
From a miniature by Lucas van Leiden in *Splendor Solis*

NOTABLE ALCHEMICAL MANUSCRIPTS

He fixed the lead until it first became hard, then fluid, and later on soft like wax. Of this prepared lead he took 20 loth [ten ounces] and one mark of pure unalloyed silver and, with a flux, heated the substances for half an hour. Thereupon he parted the silver, cast it into an ingot, when half of it was gold.

Trimosin says:

I was grieved at heart that I could not have this art, but he refused to tell his secret process and shortly afterward he tumbled down a mine and no one could tell the artifice he used.

So Trimosin continued his journey and at length reached Italy, where he found an Italian tradesman and a Jew who understood German. "These two," he says, "made English tin look like the best silver and sold it largely." They engaged Trimosin as a servant to look after the fires in their laboratory, and in this way he learned their art, "which worked with corrosive and poisonous materials." He stayed with them fourteen weeks, and then went off to Venice with the Jew, who sold to a Turkish merchant forty pounds of this silver. While the Jew was bargaining with the merchant Trimosin took six loth of the silver and brought it to a goldsmith and asked him to test it. The goldsmith directed him to an assayer of St Mark's Piazza, and he found he was a portly and wealthy man with three German assay assistants. They brought the silver he gave them to the test with strong acids, but it did not stand the test; then they spoke harshly to him, and asked him where he had got it. He told them he had come to have it tested, and to know if it was real. When Trimosin thus discovered the fraud he left the Jew and took refuge in an institution for destitute strangers.

The following day he met on St Mark's Piazza one of the German assay assistants, who asked him if he had any more of the silver. He told him that he had no more, but that he knew the art of making it and would not mind telling him.

Through this man he obtained a post with a nobleman who kept a laboratory and who wanted a German assistant. Here the chief chemist, named Tauler, engaged him at a weekly wage of two crowns and board. The laboratory, Trimosin states, was in a large mansion called Ponteleone, about six miles from Venice, and here he took up his duties. "I never saw such laboratory work in all kinds of particular processes and medicines as in that place," he declares. "Each workman had his own private room, and there was a special cook for the whole staff of laboratory assistants."

The chief chemist gave him an ore to work on which had been sent to the nobleman four days previously. It proved to be cinnabar which had been covered with all kinds of earth to test his knowledge. This kept Trimosin busy, and he states that, with the particular process he employed, he found on testing the ingot of the fixed mercury that the whole weighed nine loth, and the test gave three loth of fine gold. "This," he says, "was my first work and stroke of luck." The nobleman was delighted and, tapping him on the shoulder, called him his *fortunatum* and gave him twenty-nine crowns. Trimosin was then called upon to swear that he would not reveal his art to anyone.

The nobleman, he tells us, was a great patron of the alchemical art and a collector of manuscripts.

> I myself witnessed that he paid 6000 crowns for the manuscript *Sarlamethon*, which described a process for a 'tincture' in the Greek. This he translated and gave me to work, and I brought that process to a finish in fifteen weeks, therewith I 'tinged' three metals into fine gold, and this was kept most secret.

He continues:

> Unfortunately, this powerful and gorgeous nobleman went to the annual ceremony at Venice of throwing a gem ring into the water at the wedding of the Adriatic, and while on board a great pleasure-ship a hurricane suddenly arose, and he, with many others of the Venetian lords and rulers, was drowned. The laboratory was then closed by the family.

Afterward Trimosin had a number of Cabbalistic and other books on magic entrusted to him, some of which he translated, and at length he "captured the Treasure of the Egyptians."

NOTABLE ALCHEMICAL MANUSCRIPTS

He says that he also discovered the great subject they worked with, and that the ancient heathen kings used such 'tinctures' and operated with them. After studying these books for a long period he began to see the fundamental principles of the art and then began working out the best tincture.

When I came to the end of the work I found such a beautiful red colour as no scarlet can compare with and such a treasure as words cannot tell. One part 'tinged' 1500 parts silver into gold.

Solomon Trimosin's alchemical notes as recorded in the *Aurum Vellus*, showing the processes by means of which he made his 'tinctures,' are not without interest and have been summarized by Schmieder. He describes how to make the "Mercury of the Philosophers," or the "Mercurial Water," and also the "Lion's Blood." The latter was prepared by dissolving gold-leaf with the mercurial water in a glass retort and dividing the residue into two parts. On one half alcohol was to be poured and the mixture allowed to digest with gentle heat for fifteen days until it became red. This was called the "Lion's Blood," and when kept at the heat of the Dog Days in a hermetically sealed retort provided the first 'tincture.' One part of the red tincture was then to be wrapped in paper and projected on 1000 parts of gold in fusion, and after remaining in fusion for three-quarters of an hour, "tincture B" was produced. One part of "tincture B" projected on 1000 parts of fine silver, tin, or lead was said to transmute the base metal into fine gold.

"Tincture B" dissolved in strong alcoholic wine Trimosin regarded as the Elixir, and "if a spoonful be taken in the morning," he states, "it will strengthen and renew the constitution, rejuvenate the aged and make women prolific."

There is no evidence to prove that Solomon Trimosin ever existed or that he wrote *Splendor Solis* or the other works attributed to him. Their origin is unknown, but Schmieder, who made a careful investigation of the matter, believes that the author of *Splendor Solis* was called Pfeifer and that he was a Saxon by birth.

CHAPTER XVI

ALCHEMISTS AND ROYALTY

ALCHEMY appears to have assumed its highest importance in the fourteenth and fifteenth centuries, when the lure of the Philosopher's Stone attracted so many to its pursuit, but in the thirteenth century Alfonso X of Castile had written an alchemical treatise entitled *The Key of Wisdom*. It is little wonder that later many of the reigning kings and princes of Europe became interested in the art of gold-making in the hope of increasing their wealth.

The alchemist, owing to the powers he claimed to possess, began to play an important part in State affairs, and in the fifteenth century he had a place in nearly every Court in Europe, and often became the confidential adviser of the ruler of the country. This being an age when human credulity was easily impressed, the influence he exerted, as may be imagined, was often misused, and the fraud and imposture to which alchemy gave rise resulted in laws being made to stop its practice.

The pursuit of alchemy was regarded by the State as a possible source of revenue as early as 1330, for we find that Thomas Cary was ordered to bring before King Edward III John le Rous and Master William de Dalby, who were said to be able "to make silver by alchemy, with the instruments and other things needful to their craft."

In 1414, during the reign of Henry IV, an Act was passed forbidding the use of the craft in its efforts to multiply gold, and the penalty for contravening it was considerable. On the other hand, the practice of alchemy was legalized pursuant to letters patent, and various persons were granted permission or licences to carry on the art of transmuting metals. Thus we find that in 1444 one Richard Cope was authorized "to transmute the imperfect metals of their own kind by the art of Philosophy and to transubstantiate them into gold or silver."

Two years later licences were granted to Edmund Trafford and Sir Thomas Ashton for a like purpose, and in the same year (1446) John Faunceby, John Kirkeby, and John Rayney

LICENCE GRANTED TO JOHN ARTEC TO PRACTISE ALCHEMY AND CONTINUE HIS WORK OF TRANSMUTATION
Sixteenth century

received the royal permission to search for the Philosopher's Stone and the Elixir of Life and to transmute metals. It is supposed that the necessity for this royal licence is based on the sovereign's claim to all mines and therefore to all other sources of the precious metals.

In 1463 King Edward IV granted to Sir Henry Grey, of Codnor, in Derbyshire, the authority to "labour by the cunning of philosophy for the transmutation of metals with all things requisite to the same at his own cost, provided that he answered to the King if any profit grow therefrom." Apparently Sir Henry did not reap much profit for his royal patron, for it is recorded that two years later the King decided that he had had sufficient time for his experiments and called upon him to render an account of his gains. He did not, however, answer to the summons, so his case was postponed from term to term for five years. At length a date was fixed for him to appear at Court in the middle of October 1470, but, says the chronicler,

> before that date the Lord King, certain necessary and urgent causes moving him, made a journey from this realm of England to foreign parts leaving no regent or guardian in the same realm, wherefore the Barons of the Exchequer did not come to hear pleas.

So Sir Henry apparently escaped being called to account. Edward IV, however, still continued to have belief in alchemy, and in 1476 he licensed David Beaupee and John Merchaunt to practise for four years "the natural science of the generation of gold and silver from mercury," but the results of their experiments do not appear to have been recorded.

His predecessor, Henry VI, also had dealings with alchemists and granted several licences of a similar kind, while Henry VIII, who was more interested in medicine, has left a manuscript book of recipes recommended for various diseases, several of the formulæ having been "devysed by the King's Majestie."

Queen Elizabeth was not only a patroness of alchemists, but also a believer in the art. She had dealings with John Dee and his confederate, Edward Kelly, with regard to transmutation, and had more than one alchemist in her employ. According to a manuscript now in the British Museum and other records in the Calendar of State Papers, on February 7, 1565, Cornelius Alvetanus, otherwise Cornelius de Lannoy, was engaged by the Queen to produce 50,000 marks of pure gold annually at a moderate charge. He was allotted a room in Somerset House to use as a laboratory, and there, after writing a treatise entitled *De conficiendo divino Elixire sive lapide philosophico*, which he dedicated to Queen Elizabeth, he began his operations in making gold.

The Princess Cecilia, daughter of Gustavus I of Sweden, was at this time living as an exile in London and lodged near Somerset

ALCHEMISTS AND ROYALTY

House, where she sought out de Lannoy. She was heavily in debt and entered into negotiations with the alchemist, with the result that he signed a bond on January 20, 1566, to lend the Princess the sum of £10,000 on the following May 1, to be repaid in thirteen yearly instalments of £1000 each, in consideration of a payment of £300 to de Lannoy. By some means the story of this transaction came to the ears of the Queen, and she was highly incensed that others were attempting to reap the benefit of her protection. She at once forbade de Lannoy to hold any further communication with the Princess, meanwhile placing him under observation. De Lannoy, aware that he had aroused the Queen's suspicions, wrote to Sir William Cecil, stating that "our great and glorious design has fallen into suspicion, but I swear on the Holy Gospels that I will carry it through successfully and promise to hold no communication with the Princess." He, however, proved faithless to his protestations and promise, and on Cecil hearing that he was preparing to leave England with the Princess he was at once arrested and lodged in the Tower.

In July 1566 the discomfited alchemist addressed a letter to the Queen in which he said:

> I know how grievous this delay must be to you. I have nothing to offer you in this kingdom but my life, which would be a heavy loss to my innocent wife. As to the business of transmuting metals and gems to greater perfection, either the work has been disturbed or some wicked man has been present or I have erred through syncopation. Pray permit me to write to my friends for help, for I can indubitably perform what I have promised.

On August 3, 1566, he made a declaration that, if it should please the Queen to release him from confinement, he would without delay put into operation that wonderful elixir for making gold for her Majesty's service. Apparently no notice was taken of this appeal, for on August 13 he wrote from the Tower of London to both the Earl of Leicester and Cecil, asking them to intercede for him and imploring the Queen's mercy.

Apparently their intercession was successful, for thirteen days later Sir F. Jobson and Armigill Waad wrote to Cecil saying that they had conferred with Cornelius and that requisitions were being made by him for carrying on his alchemical operations, for which a little money would be required. Whether the unfortunate alchemist was actually liberated or not we are not told, but Cecil records in his diary in January 1567 that "Cornelius de la Noye was sent to the Tower for abusing the Queen's Majesty in Somerset House in promising to make the Elixir," and again on February 10 of the same year he wrote that de Lannoy abused many in promising to convert any metal into gold.

The last reference to de Lannoy among the State Papers appears to be a letter which he wrote to Cecil on March 13, 1567, in which he again promises to perform the things mentioned in his offers to the Queen and solemnly engages to produce gold and gems by a chemical process. He may have ended his days in the Tower, for we hear nothing further of him or his wonderful process for making gold.

In spite of this and other disappointments Queen Elizabeth continued to have faith in alchemy, for in a document dated February 20, 1594, we find the following instructions to

> Robt Smith of Yarmouth sent by the Queen to Lubec. He having received the Queen's reply to a letter from Ruloff Peterson of Lubec, is to repair thither, deliver the letter and receive the three glass bodies and bring them to her Majesty. He is to ascertain from Peterson whether the materials therein were considered by Ouldfield to be brought to full perfection, and if anything is lacking what is it? Also, to recover any books or papers of Ouldfield relating thereto, or other of his books which treat of alchemy; also a secret menstruum, without which the materials aforesaid can hardly be brought to perfection. All these things are to be brought to her Majesty in order to ascertain their value and either detain them or return them on the consideration mentioned (*viz.*, £500) if she kept them.

What mysterious operation was here involved we can only conjecture, but the reference to the "secret menstruum" doubtless

ALCHEMISTS AND ROYALTY

indicated the Philosopher's Stone, for which Elizabeth was apparently willing to risk £500!

Dr John Dee, the astrologer and mathematician, records in his diary that "E.K[elly] made projection with his powder in the proportion of one minim upon an ounce and a quarter of mercury and produced nearly an ounce of best gold." The news of Kelly's success reached the ears of Lord Burleigh, who wrote to him for "a specimen of his marvellous art," and it was afterward reported that a warming-pan from the copper or brass lid of which a piece had been cut and transmuted into gold and replaced was sent to the Queen. Even Elias Ashmole was deceived by Kelly's tricks, for he writes that from "a very credible person (who had seen them) Kelly made rings of gold wire twisted twice round the finger which he gave away to the value of £4000."

The lure of the Philosopher's Stone also drew many high dignitaries of the Church in the quest for gold, and among them was Sir Thomas Ellis, who was Prior of Leighs, in Essex. He was reputed to have had such skill in transmuting metals that he was suspected of coining, and was at last called upon by the authorities to give an account of how he came to practice alchemy. He told them that he had become interested in the art by reading of it in books, and, being desirous of acquiring a knowledge of the science, he had entered into communication with one Crawthorne, a goldsmith who lived in Lombard Street, London, who introduced him to a priest called Sir George, who was said to be cunning in these matters. This priest introduced him to Thomas Peter, a clothworker, who "sayed he had the syens of alkemy as well as eny man in Yngland." The prior promised to pay Peter £20 for instruction in the art and gave him twenty nobles in advance; whereupon Peter gave him some silver and quicksilver with directions how to treat them. They were to be sealed hermetically in a glass vessel, which was to be placed in an earthen pot of water which had to be kept hot for ten weeks or more. Sir Thomas Ellis says:

ALCHEMY AND ALCHEMISTS

Master Peter came from time to time to see how matters were progressing with the prior's experiment, but before the ten weeks were up he managed to break the glass vessel. The prior, in disgust, then sold the silver for what it would fetch and refused to pay Peter the balance of the £20. The latter threatened an action for the debt, but as it chanced that twenty marks was paid at this time to the prior for the lease of a rectory, he handed the money over to Master Peter to settle his claim. I never medelyd with him syne nor with the crafte nor never wyll. God wyllyng.

Charles II had a well-equipped laboratory in his palace at Whitehall, where he carried on experiments in chemistry, in which he was much interested. Burnet alludes to it in 1685, and states that when the King was unable to walk he spent much of his time in his laboratory and was running a process for fixing mercury. Pepys in his *Diary* mentions on January 15, 1668, that he went to see "the King's Elaboratory underneath his closet. A pretty place, and there saw a great many chemical glasses and things, but understood none of them."

Charles was also interested in several medicinal preparations, some of which he made in his own laboratory. One of these, which acquired a considerable reputation, was known as 'King Charles' Drops.' Dr Martin Lister, in referring to it in 1694, says:

> The late King Charles not only communicated to me the process, but very obligingly showed it to me himself by taking me alone into his elaboratory at Whitehall while the distillation was going on. Mr Chevins on another occasion showed me the materials for the drops which were newly brought in—*viz.*, raw silk in great quantity. One pound of raw silk yielded an incredible quantity of volatile salt and in proportion the finest spirit I ever tasted, and what recommends it is that when rectified it is of far more pleasant smell than that which comes from sal ammoniac or hartshorn, while the salt refined with any well-scented chemical oil makes the 'King's Salt,' as it used to be called.

The versatile Prince Rupert was also an enthusiastic student of chemistry, and in a manuscript now in the British Museum which is attributed to him are recorded several interesting

ALCHEMISTS AND ROYALTY

processes. They include formulæ for making "the regulus of antomony, crocus martis, and a process to blanch and sublime sal ammoniac." Attached to the manuscript is the following letter, apparently addressed to his instructor:

> I intreate you to do me a favour because my glasse pott is broken in which I melt my medicines and till I get another I am destitute.
>
> Your friend and student

Prince Augustus of Saxony attained quite a reputation as an alchemist during the second half of the sixteenth century. One of his assistants was David Benter, who on failing to produce gold to satisfy his master was thrown into prison at Leipzig. His endeavours to escape proving unsuccessful, he inscribed on the walls of his cells, "Caged cats catch no mice," which is said to have so amused the Prince that, after Benter had renewed his promises to make gold, he was set at liberty so that he could renew his operations. But again he was unsuccessful, so he poisoned himself, thus ending his days without accomplishing his task.

The Prince worked in his laboratory at Dresden, which was known to the citizens as the "Gold House," and here, in 1577, he believed he had discovered the great secret, for he wrote to Francesco Forense, the Italian alchemist, "I have now reached such perfection in transmutation that I can make daily three ounces of good gold from eight ounces of silver." But the Prince in his enthusiasm had evidently deceived himself, for in 1585 he engaged the services of Sebald Schwertzer, who claimed to have found the tincture and demonstrated his ability before the Elector by transmuting three marks of quicksilver into gold.

The Director of the Treasury calculated that the tincture had transmuted 1024 times its weight of metal, and Schwertzer undertook to make ten marks of gold daily, but the death of the Elector put an end to the operations. On leaving Dresden

Schwertzer went to Prague, which at that time was the Mecca of the alchemists, and there was cordially received by the Emperor Rudolph, who eventually appointed him director of the Imperial mines at Joachimsthal.

The most famous of all the royal alchemists of the sixteenth century was the Emperor Rudolph II of Germany, who was born in Vienna in 1552. On the death of his father in 1576 he succeeded to the throne and became King of Hungary and Bohemia and went to live in the great Hradschiner Palace at Prague. Deeply interested in the occult sciences, he became an enthusiastic student of alchemy and astrology, and so attracted to his Court from all parts of Europe the men who professed to practise those arts in order to secure his patronage.

Most of these pseudo-alchemists were mere cunning pretenders who travelled from country to country in search of credulous patrons or any who would lend an ear to their romantic stories. It was chiefly men of this type who flocked to Prague in the hope of enlisting the sympathy and help of the Emperor Rudolph, whose pursuits will be described in the following chapter.

The Emperor Ferdinand III is said to have purchased the secret of making the Philosopher's Stone from Richthausen, of Vienna, who is said to have transmuted three pounds of mercury into two and a half pounds of gold by means of his red powder on January 15, 1648. A medal was made from this gold which was preserved in the Treasury at Vienna, and Richthausen was made a baron. (See page 157.)

Among other royal alchemists Leopold I and Frederick William I and his successor, Frederick William II, Kings of Prussia, were firm believers in alchemy; several of the Kings of France were also deeply interested in the art, including Charles VI, Charles VII, and Charles IX. The first-named was one of the most credulous monarchs of his day, and his Court is said to have swarmed with alchemists, astrologers, and charlatans of every description. The King himself made several attempts to

ALCHEMISTS AND ROYALTY

discover the Stone, and wrote a treatise called *Royal Work of Charles VI of France and the Treasure of Philosophy*. Charles VII is said, when at war with England, to have been enticed by an alchemist named Le Cor into the production of large quantities of counterfeit gold, with which, in the form of coinage, he inundated the neighbouring countries of Europe. Christian IV of Denmark and Charles XII of Sweden did not disdain to enlist the aid of the alchemists of their times to fill their treasuries by means of the Philosopher's Stone.

Among the prisoners of war taken by the troops of Charles XII at Warsaw in 1705 was a Saxon lieutenant named Paykull, who claimed to be an adept in alchemy. He was condemned to death, but the King promised him a respite, the condition being that he should make two million dollars' worth of gold each year. To this Paykull agreed, and he is said to have transmuted lead into gold by means of a 'tincture' which he had made from a secret formula. In the presence of Hamilton, the Master of Ordnance, he is also stated to have changed six ounces of lead into gold which was declared to be worth 147 ducats. From this gold two medals were struck which were inscribed, "Hoc aurum arte chemica conflavit Holmiæ. 1706. O. A. v. Paykull." In spite of his endeavours and his alleged success, the sentence upon Paykull was eventually carried out.

The Cardinal Prince de Rohan was one of the last great Church dignitaries to be bitten with the craze for alchemy. About 1780 he equipped and furnished a fine laboratory at his magnificent palace at Saverne, where he entertained Cagliostro in splendid style. Here they carried on their operations with great secrecy and apparent success, for the Cardinal afterward declared to the Baroness d'Oberkirch that "Cagliostro had made not less than five or six thousand livres of gold."

CHAPTER XVII

ALCHEMISTS IN FORTUNE AND MISFORTUNE

ACCORDING to Adam von Bremen, there was an official alchemist appointed to the Court of Adalbert in Germany as early as 1063. He was a baptized Jew named Paul, who gave out that he had learned in Greece the art of transmuting copper into gold. There were other Courts in the Middle Ages where alchemists held important positions, such as that of Frederick, Duke of Würtemberg, and of Anna of Denmark, who was herself an ardent seeker of the Stone and built two laboratories on her estate in which it is said "furnaces great and small were kept glowing night and day." But in the sixteenth century the Imperial Court of the Emperor Rudolph II at Prague surpassed all others as a centre of attraction for practitioners of the occult arts.

The Emperor's laboratory in the old city, Bolton tells us, was housed in a one-storied building containing two rooms which communicated with each other. The floors were flagged, and on one side were several flues into which the brick furnaces, arranged along the wall, discharged their fumes. One of these was used for smelting ores, another for producing moderate heat for a great water-bath, and a third for the distillation of volatile liquids. This furnace supported a cucurbit capped by five helms placed one above the other, their long necks terminating in recipients for collecting the distillates. On the shelf-lined walls stood an array of cucurbits, alembics, descensories, matrasses, and pelicans, together with glass jars, bottles, and gallipots, containing chemicals solid and liquid. On a wooden block stood a great mortar, its pestle attached to a spring-beam

FORTUNE AND MISFORTUNE

fastened to the smoke-begrimed rafters. Near a window on a ledge lay books and manuscripts for reference and study, while a table near by was littered with flasks, funnels, sand-glasses, and knives. In the corridor between the laboratories lay heaps of charcoal, crucibles in nests, boxes of materials for lutes, and utensils of iron, copper, and brass. Here worked such adepts as were in favour, and sometimes their royal patron would pay them a visit to watch their operations.

In the sixteenth century Prague retained the features of a medieval city, with its narrow, winding alleys and ghetto. In the Hradschin quarter, near the cloisters of St George's Church, ran a tortuous, steep street that was known as Gold Alley, for here lodged many of the alchemists and other adventurers who had been attracted to Prague. Among them may be mentioned Daniel Prandtner, Christopher von Hirschenberg, Master Jeremias, Leonhard Vychperger von Erbach, and Michael Maier the mystic. The last-named eventually became one of the secretaries to the Emperor and later was an exponent of the Fraternity of the Rosy Cross. Here also dwelt Claudius Syrrus, an Italian alchemist in the service of Prince von Rosenberg, who had made a contract with his master to discover the secret of transmutation.

Among those who lived in or near the royal palace was Dr Thaddeus von Hayck, a personage of great importance, who held the office of Court physician and was director of the alchemical laboratories. To him came those who sought the Emperor's favour and patronage, for von Hayck had great influence with his royal master.

Among the visitors to his apartment one day were Dr John Dee and Edward Kelly, who had journeyed from England to give demonstrations in proof of their spagyric powers by the aid of the "shew stone." It was in the cellar laboratory in von Hayck's house that Kelly also gave a demonstration of his skill in transmutation, and succeeded in deceiving his learned host so effectively that he was thanked by the doctor, who told him that as director of the laboratories of the Imperial Court he had detected many kinds of frauds. Some impostors, he said, had used double-bottomed crucibles, the false bottom being made of powdered crucible, clay, and earth, mixed with wax, gold filings being concealed in the space between. Another adept he had detected in dropping into the crucible a piece of charcoal in which gold-leaf had been hidden. Those who pretended to make gold and silver without fire used *aqua fortis* in which silver had been first dissolved. Others had a knife-blade made of two metals soldered together, the golden half painted black with a varnish soluble in alcohol; the removal of the coating from the gold by immersion in spirit of wine effected the deception. He also told his visitors of another trick, bleaching copper with a preparation of arsenic, and that the most common fraud of all consisted in using an amalgam of gold and mercury, from which when heated the mercury evaporated and left the precious metal behind. Although the recounting of these exposures must have made Kelly feel very uncomfortable, Dee's success with his "shew stone" caused him to be received by the Emperor, with whom he had a long audience. This was followed by conversations at which hermetic philosophy and the Emerald Table were discussed. It was arranged at length that Dee and Kelly should give a demonstration with the "shew stone" before the Emperor and exhibit their powers. During the *séance* Kelly professed to give, through the medium of the spirit Zadkiel, a formula for making the Philosopher's Stone which was taken down by the Emperor for further investigation, and for a time the two alchemists basked in the royal favour and lived in luxury. After a while, however, Dee began to meddle in Court intrigues and fell into disgrace, which resulted in their hurried departure from the city.

Some of the adventurers at Prague were not without a sense of humour, judging from the story of Mardocheus de Delle, the Court poet, who one day announced that Benedict Töpfer had discovered that

RUDOLPH II. EMPEROR OF GERMANY, IN THE LABORATORY OF HIS ALCHEMIST
Vaczlar Brozik
By courtesy of the New York Public Library

THE ALCHEMIST
From the painting by Giovanni Stradano in the Palazzo Vecchio, Florence
Photo Alinari

FORTUNE AND MISFORTUNE

gold could be made out of Jews. He had found by experiment that twenty-four Jews yielded by proper treatment half an ounce of gold, so that by repeating the process daily with a hundred Jews, making due allowance for Holy Days, 624 ounces of gold could be made in twelve months.

Other amusing stories have survived showing how even the charlatans themselves were duped. It is said that one day an imposing Arab, in gorgeous robes with turban complete, arrived in Prague and established himself in a fine house, where he lived in princely style. He appeared to have ample means and soon became known to the many alchemists and adventurers in the city, seeking their company and friendship. At length he invited twenty-four of them to a banquet at his house, and after the feast, at which a large quantity of wine had been consumed, he proposed to his guests an experiment in which he declared that he would demonstrate to them a method of multiplying gold in which all could participate. Every one who contributed a hundred marks would certainly receive a thousand, and he assured them that the operation was perfectly safe. Nearly all agreed, some sending to their houses for the necessary money, and others, who had it, lending to some who had not sufficient.

The Arabian host, having collected the gold and added his own contribution, led the way to his laboratory. His guests noticed that it was well equipped with furnaces already glowing, retorts, aludels, and apparatus of every description. Carefully selecting a very large crucible, into which he apparently placed all the golden coins, he added some *aqua fortis*, copperas, mercury, lead, salt, eggshells, and dung, and placed the vessel in one of the furnaces.

The eager guests crowded round, watching him with suppressed excitement. He seized the handle of the bellows ready to blow, when suddenly there was a terrific explosion which scattered the live coals and filled the laboratory with suffocating gases that attacked all present. Some were burned by the fiery embers, others were nearly asphyxiated by the fumes, and utter darkness added to the confusion.

When lights were at length obtained the more venturesome searched for the host to see if he was injured, but there was no trace of him. The Arabian alchemist had disappeared in the smoke, and all they could discover was an open window leading to an alley, the fragments of the broken crucible, the ruins of the furnace, and a mass of broken apparatus. In his rapid flight the alchemist did not forget to take with him the 2400 marks which he had collected from the pockets of his guests, and he was never seen in Prague again.

During the summer of 1590 another mysterious individual made his appearance in the city and excited the curiosity of the inhabitants. He was said to be a wealthy Italian who had discovered the great secret. His name was Alessandro Scotta. He rented a splendidly furnished house and spent his money so lavishly that most of the nobles and important people at the Court hastened to call upon him and offer him hospitality. He rode through the streets in a magnificent carriage lined with crimson velvet, and this gorgeous equipage, with outriders, was followed by three others filled with his attendants. The procession wound up with an armed bodyguard. All this display attracted great attention. His name naturally spread abroad in the city, and he soon got an introduction to the Emperor, who offered him the use of a laboratory to carry on his operations, but after a while, being unable to produce any results, he suddenly left Prague for Coburg, where he managed to dupe the Duchess with the story of his discovery of the Stone and then fled to his native country.

Another of the pseudo-alchemists who found his way to Prague was an Italian, one Mamugna, who styled himself Count Marco Bragadino, and declared that he was a Greek. This impostor first sought fame in Vienna, where he asserted that he had discovered the secret of transmutation. When in Prague he was always accompanied in the streets by two large black

FORTUNE AND MISFORTUNE

mastiffs who were supposed to be his familiar spirits. On leaving Prague he travelled to Munich, where he procured an introduction to the Duke of Bavaria and succeeded in obtaining from him a large sum of money on the credit of his story that he was in possession of the secret, but subsequently he was detected in his fraud, arrested, and sentenced to death in 1591. The authorities staged a picturesque ending to the pseudo-alchemist, for he was brought to the gallows clad in gilt tinsel and hanged with a yellow rope, his two black dogs being killed at the foot of the gibbet and thrown into the same grave as their master.

Thus many of these fraudulent practitioners, who travelled from one country to another and lived in luxury for a time by duping their wealthy and credulous patrons, eventually paid the penalty and ended their days in prison or were put to death, some after suffering horrible tortures.

Among these unfortunates was a woman named Marie Ziegler, who, on failing to fulfil her promise to give Duke Julius of Brunswick a recipe for transmuting base metals into gold, was sentenced to be roasted alive in an iron chair, which sentence was carried out in 1575. George Honnauer, who in 1597 promised the Prince of Würtemberg to transmute thirty-six hundredweight of iron into gold, was detected by a boy who had been concealed in the laboratory in the act of putting gold into the crucibles. He was arrested and was sentenced to be hanged on a gallows made of iron. Nine years later another impostor, named Andreas von Muehlendorf, was executed at Stuttgart on the same gallows for failing to carry out a similar contract. About 1677 Colonel Krohnemann, who was in the service of the Margrave of Brandenburg, claimed to have been successful in transmuting base metals into gold and was subsequently appointed Director of the Mint and Mines. In proof of his statement he had seven medals struck from the gold he declared he had made, but the last one, which he dedicated to the Margravess Sophia Louise of Brandenburg, proved his undoing, for

ALCHEMY AND ALCHEMISTS

he was suspected of deception and imprisoned in the citadel of Plassenburg in 1681. Here he obtained permission to continue his experiments, which he carried on for five years; then he managed to escape. He was, however, recaptured, tried, and found guilty of having used gold from the Margrave's treasury for his operations. The formula from which he claimed to have produced his gold was made public at the time of his trial and was as follows:

> Mercury, verdigris, vitriol, and salt are to be digested with strong vinegar in an iron pot and stirred with an iron rod until the mass takes on the consistency of butter. The remaining liquid, which is an amalgam of copper, is then to be pressed through leather and put into a crucible with even parts of curcuma and tutia, whereupon the crucible is to be heated by a blast. The curcuma reduces the tutia (impure oxide of zinc), and the copper in the amalgam unites with the zinc to form brass.

It was found that Krohnemann had added gold to this alloy of copper and zinc in sufficient quantity to answer the tests and so carried out his deception.

Another alchemist of this type, who, however, escaped the extreme penalty for his frauds, was Leonhard Thurneisser, the son of a Swedish goldsmith. He set out on his adventures while a young man, and travelled from country to country, adding to his knowledge of alchemy and medicine whenever he had the chance. On reaching Germany he managed to secure the confidence of the Archduke Ferdinand, who supplied him with money and sent him to travel in the East. On his return he entered the service of the Elector of Brandenburg and was appointed the director of the laboratory built by his wife. He commenced to practise medicine in Berlin and was at first very popular, but his success excited the envy of the regular physicians, who denounced him as a quack, and in the end he left the city in haste. We next hear of him in Rome, where he was received by various notables, including the Cardinal Fernando di' Medici, who invited him to dine with him. After dinner

FORTUNE AND MISFORTUNE

the alchemist offered to entertain the distinguished company gathered to meet him by transmuting half an iron nail into gold. The guests, of course, were delighted. Thurneisser produced an ordinary iron nail, and, after carefully warming it, dipped it into an oily liquid which he had with him. On his withdrawing it one half of the nail, to their astonishment, appeared to have been changed into gold.

This was an old trick of the pseudo-alchemists which they performed with a nail made partly of gold and partly of iron, from which a solvent removed a black coating and disclosed the yellow metal. However, he convinced his audience, for the famous nail was long preserved in the palace of his Eminence, accompanied by a certificate signed by the Cardinal and dated at Rome November 20, 1586.

In spite of his cleverness Thurneisser fell on evil days in Italy, and is said to have died in poverty in a monastery.

Richthausen, of Vienna, who is said to have sold the secret of the Philosopher's Stone, which he claimed to possess, to the Emperor Ferdinand III, is stated to have acquired it from an adept named Busardier, who lived at Prague. While on his deathbed Busardier sent Richthausen a message that he wished to see him so that he might hand the secret over to him. Richthausen at once set off, but on arriving at Prague he found that the alchemist was dead. He managed, however, to obtain some of the 'powder of projection,' after which he immediately left the city, but was pursued by the steward of Busardier's noble patron, who, pointing a pistol at his head, demanded its restoration. Richthausen, in fear for his life, handed him a packet containing a small portion, but himself retained the greater part of the precious powder.

Later on Richthausen returned to Prague and presented himself at the Court. The Emperor, interested in his story, called in the aid of Count Russe, his Master of Mines, and it was arranged for Richthausen to give a demonstration of his skill. According to the account, with a single grain of his powder he succeeded in transmuting three pounds of mercury into gold, and the Emperor was so delighted that he had a medal struck from the metal to commemorate the event. It was inscribed, "Divina metamorphosis exhibita. Prague Jan. 15. Anno 1648 in presentia Sac. Cæs. Majest. Ferdinandi Tertii." Richthausen was afterward created a baron, and in 1651 made a further demonstration at Mayence in the presence of the Elector of that city. A more detailed account of this operation is recorded, and it is thus described:

> A small quantity of the powder was enclosed in gum tragacanth and put into the wax of a taper, which was lighted and placed in the bottom of the crucible. These preparations were carried out by the Elector himself, who then poured four ounces of quicksilver into the crucible and placed it on a charcoal fire.
>
> By blowing and raking the coals, in thirty minutes the contents were over red; the proper colour being green. Then the Baron said the matter was "yet too high and that silver must be added."
>
> The Elector then threw in some coins, which melted and combined with the contents of the crucible. When all was fused it was poured into a 'lingot' and after cooling proved to be fine gold but rather hard. On a second melting it became exceedingly soft, and the Master of the Mint declared to the Elector that it was more than twenty-four carats and that he had never seen so fine a quality of the precious metal.

Although Richthausen, like many others of his kind, at first succeeded in duping his Imperial and noble patrons, his frauds were eventually discovered, and he ended his days in disgrace.

CHAPTER XVIII

FAMOUS ALCHEMISTS OF THE SIXTEENTH CENTURY

AMONG the alchemists who became famous in the sixteenth century the names of Basil Valentine, Paracelsus, and Cornelius Agrippa stand prominent. The study of alchemy at this period attracted many men of great intelligence, with the result that a real advance was made in the science. Some of them combined the practice of alchemy with that of medicine and began to apply it to the treatment of bodily ailments and diseases, employing the chemical substances they had discovered, which they justified by their belief that the vital processes of the human body were chemical in their nature.

The identity of Basil Valentine is a mystery that has never been satisfactorily solved. Whether he was a Benedictine monk who lived at Erfurt or at Walkenried, or whether a real person of this name ever existed, has never been determined. The archives of the Benedictines make no mention of his name. He is supposed to have flourished in the early part of the sixteenth century, but the date and authority of his works have long been disputed. They are said to have been originally circulated in manuscript, but no copies are known to exist. Some declare them to be forgeries dating from about the year 1600, the information having been culled from various writers, while others declare them to be the work of Johann Thölde, a metallurgist and owner of some salt-mines at Frankenhausen, in Thuringia, who published them and used the name 'Valentine' as a pseudonym. In any case, the author of the works we know as Basil Valentine's is shown to have been the most distinguished alchemist of the period and one who had a remarkable knowledge of the science. The several treatises bearing his name were not printed until the second half of the seventeenth century, and the one by which he is best remembered, *The Triumphal Chariot of Antimony*, was not known until 1685. In this work he records many experiments made with antimony and describes its medicinal properties. He made the regulus of this metal, together with its 'glass,' an oxysulphide, and an oil, an elixir, flowers, liver, and a balsam.

He boldly advocated the use of chemical preparations for medicinal purposes, which led to the iatro-chemical period, inaugurated by Paracelsus, that later dominated medicine. He shows a remarkable knowledge of practical chemistry, and appears to have had some idea of qualitative analysis, as he was able to detect several metals when mixed in small traces with others. He shrewdly observes that much of the apparent transmutation carried on by the fraudulent alchemists was due to the effect of mixing metals in different proportions, and that these mixtures contained neither gold nor silver. In other treatises under his name, such as the *Revelation of the Hidden Key* and *The Great Stone of the Philosophers*, he describes how to obtain spirit of salt, the action of oil of vitriol on sea-salt and brandy distilled from wine. He obtained copper from pyrites by first obtaining a sulphate and then precipitating the metal by plunging into the solution a bar of iron. He was acquainted with arsenic, zinc, bismuth, and manganese, and also with preparations of mercury and lead. He alludes to fulminating gold, to the double chloride of iron and ammonia, and describes methods of preparing many other metallic salts.

One of the most disquieting considerations for those who seek to identify Basil Valentine with a monk of the sixteenth century is that in his works we find facts which are far in advance of those generally known at that time, and yet with them we have his adherence to the ancient alchemical theories. It is now generally believed that his works were written after the time of

ALCHEMISTS OF THE XVIth CENTURY

Paracelsus. He enunciates the same ideas regarding processes happening within the earth as those pronounced in the Emerald Tablet—*viz.*, "Cause that which is above to be below," and "Let that which is below become that which is above." Gold was the intention of nature in regard to all metals. Plants were preserved by their seed; therefore there must be seed in metals, which is their essence. If the seed could be separated and brought into a proper condition it could be caused to grow into the perfect metal. Health and life, he contended, were to be preserved and ensured by the Philosopher's Stone, which was to be used as a universal medicine.

In *The Great Stone of the Philosophers* he states:

> Whosoever thou art that presumest to dive into the fountain of our work and hopest to obtain by my ambitious enterprise the reward of Art, I tell thee by the Eternal Creator, for a truth of all truths, that if there be a metalick soul, a metalick spirit, and a metalick body that there must be a metalick Mercury, a metalick sulphur, and a metalick salt which can of necessity produce no other than a perfect metaline body. If you do not understand this that you ought to understand, you are not adepted for Philosophy or God concealeth it from thee.

The method of preparing the Great Philosopher's Stone he reveals at the close of the book; we need give but an outline of it here. It begins:

> Proceed in the name of the Lord to the work itself.
> Take of the very best gold you can have, one part, and of good Hungarian antimony six parts. Melt these together upon a fire and pour it out into such a pot as the goldsmiths use, and when you have poured it out it becomes a Regulus.
> This Regulus is again melted and the antimony separated from it. Then add mercury and melt it again. Do this three times. Then beat the gold very thin and make an amalgam with more quicksilver.
> Let the quicksilver fume away over a gentle fire that nothing remains but the gold. Then take one part of saltpetre and like quantity of sal ammoniac and half as much of pebbles well washed. Mix them and put into an earthen retort and distill in a furnace.

The water obtained is mixed with the prepared calx of gold and water added. Then digest it in warm ashes and keep it at a gentle heat for fourteen days. Water is then added, and it is again distilled and redistilled until the gold comes over.

To this spiritualized solution of gold, rain water is added, and three parts of mercury. Then decant the water and dry the amalgam. Drive off the quicksilver, and there will remain a very fair powder of a purple colour.

Then must be made the Tartar of the Philosophers from the ashes of the vine and make a strong Lee with it to coagulation, and there remains a reddish matter, and this dissolve in spirit of wine. Then take the other part of mercury of pure gold and pour this on it and distil. The precipitated mercury and the oyl of gold are then to be mixed and placed in a glass and hermetically sealed and put into a threefold furnace and allowed to putrify for a month and become quite black.

Increase the fire and the blackness will vanish, and it changes into many colours. Increase the fire to the fourth degree and the glass will look like silver. Increase the fire to the fifth degree and it becomes like gold. Continue this and you will see your matter lye beneath like a brown oyl which at length becomes dry like granite.

He that obtains this may render thanks to God, for poverty will forsake him. For this noble medicine is such a stone to which nothing in the world may be compared for virtue, riches, and power.

He goes on to say that

> if this medicine after being fermented with other pure gold doth likewise tinge [dye] many thousand parts of all other metals into very good gold, such gold likewise becometh a penetrat medicine that one part of it doth tinge and transmute a thousand parts of other metals and much more beyond belief into perfect gold.

Valentine's process apparently consisted in coating—or, as he calls it, tingeing—the baser metals with a gold amalgam and so giving them the appearance of the precious metal, and by repeating this process he thought the whole of the base metal might be converted into gold. We find that this idea was exploited by the pseudo-alchemists at a later date, and that by its means they succeeded in duping many of their patrons.

ALCHEMISTS OF THE XVIth CENTURY

Few of the alchemists had a more adventurous and eventful career than Henry Cornelius Agrippa, who was born in Cologne in 1486. He first served as a soldier under the Emperor of Germany, and, after returning from a diplomatic mission, he began to study alchemy and eventually became a reformer and mystic. He set out to travel and journeyed through France,

HENRY CORNELIUS AGRIPPA

Spain, and Italy, and settled for a time at Pavia, where he studied medicine and became a professor at the university of that city. In his principal work, *On Occult Philosophy*, he expresses his belief in the doctrines of astrology and in the theory that the spirit of the world exists in the body of the world as the human spirit exists in the body of man. This spirit, he contends, abounds also in the celestial bodies and descends in the rays of the stars, so that things influenced by their rays become conformable to them. By this spirit every occult property is conveyed into stones, metals, herbs, and animals through the sun, moon, and planets, and through the stars higher than the planets. He was a firm believer in the efficacy of charms, which, he says, may "be worn on the body bound to any part of it or hung round the neck, changing sickness into health or health into sickness." He also recommends that they be worn in the form of finger-rings, and says:

> When any star ascends fortunately, take a stone and herb that are under that star, make a ring of the metal that is congruous therewith, and in that fix the stone with the herb under it. We read in *Philostratus Iarchus* that a wise prince of the Indies gave seven rings made after this manner, marked with the names and virtues of the seven planets to Apollonius, of which rings he wore every day one, distinguishing them according to the names of the days, and by the benefit of them lived one hundred and thirty years and always retained the beauty of youth.

In 1510 Agrippa relinquished his alchemical work for a time to resume service at Court, and, it is said, was sent by Louis XII of France and Maximilian I of Austria on a secret mission to London. He has left an interesting description of London at the time of his visit. He tells us that there was but one bridge across the Thames. Fleet Ditch had just been dredged and was navigable for large boats laden with fish and fuel up to Holborn. There was no pavement in Holborn Street, which led by the Bishop of Ely's palace and strawberry-beds, skirting the country, to the open Oxford Road, and so away, passing the hamlet of St Giles. Chancery Lane, Fetter Lane, and Shoe Lane were unpaved and in a scarcely passable condition. The city had its walls and gates; the cross in Westcheap was its newest ornament. Stepney was still a town by itself, remarkable for the pleasantness of its situation and the beauty of its scenery, and chosen, therefore, as place of residence by many persons of distinction. It was here, at the house of John Colet, the Dean of St Paul's, that he was lodged, and he remarks that his host at that time was engaged over the refoundation of St Paul's School.

ALCHEMISTS OF THE XVIth CENTURY

On the conclusion of his mission he returned to Germany and took up his residence at Metz, where he became the Town Advocate and Orator. During his term of office a serious epidemic of plague broke out in the city, and he set to work to discover preventives and remedies to combat the disease.

Returning to Cologne, he again took up his alchemical studies, but on the death of his wife he removed to Geneva, where he practised medicine and joined the Reformers. Here he wrote a work on the *Vanity of Sciences and Arts*, in which he appears to have embodied his considered judgment from his experience in the practice of medicine and alchemy throughout his lifetime. With caustic satire he describes the foibles of the physicians of the time, and alludes to their pomps and vanities and the way these "bring practice to the man with a velvet coat and rings, with certain shows of religion, who is addicted to uncompromising self-assertion." The book embroiled him in many disputes, and he made numerous enemies. He further remarks on the use of costly drugs obtained from a distance, such as scammony, which could rarely be got except in the most adulterated state, while the simples of the country, which could be prepared when they were wanted, were despised and rejected.

He charges the apothecaries with dealing in adulterated drugs, and taxes them with a vanity which drove them to cause the sick even to eat human flesh spiced, which they call 'mummy.' Surgery, he declares, is a surer science, but of an evil origin, for it was bred in war. But his most scathing criticisms are reserved for alchemy. "I pass on," he says,

> to the crucible of alchemy, which consumes not less treasure than the flesh-pots. The alchemist may earn a scanty livelihood by the production of medicaments or cosmetics, or he may use his art, as very many do, to carry on the business of a coiner. But the true searcher after the Stone which is to metamorphose all base metal into gold converts only farms, goods, and patrimonies into ashes and smoke. When he expects the reward of his labours, births of gold, youth, and immortality, after all his time and expense, at length old, ragged, rich only in misery, and so miserable

that he will sell his soul for three farthings, he falls upon ill courses as counterfeiting of money.

He adds:

> I do not deny that to this art many excellent inventions owe their origin. Hence we have the discovery of azure, cinnabar, minium, purple, that which is called musical gold, and other colours. Hence we derive the knowledge of brass and mixed metals, solders, tests, and precipitants.

Judging from his experiences related in the *Vanity of Sciences and Arts*, Agrippa appears to have tried all these things and found them wanting. During his short and eventful life—for he died before he reached the age of fifty—he had been a soldier in Germany, a professor in Italy, a knight of the Empire, secretary to Maximilian I, councillor to Charles V, a courtier in Austria, a theologian at Dôle, a lawyer at Metz, a physician in Switzerland, and as deep a searcher as any man of his day into the philosophy of the ancients. He bitterly condemns the uncertainties and vanities of the imperfect arts and sciences of the time, but, in spite of his superstitious beliefs, he was a man of real erudition.

Toward the close of his life he took up his residence at Ghent, and shortly afterward went to Brussels, where he was arrested for debt. His plain speaking in his last book made enemies of his most powerful friends, and exposed him to the vengeance of offended priests and courtiers. He died at the age of forty-nine in France while on his way to Lyons to publish some of his works.

Contemporary with Agrippa was Philippus Theophrastus Bombastus von Hohenheim, better known as Paracelsus, who was born at Einsiedeln, in Switzerland, in 1493. He was the son of a physician and as a youth evinced a love for the art of medicine, and after spending some time at the University of Basle he led a wandering life, travelling from country to country gathering knowledge and experience. Working in

PARACELSUS (1493–1541)
From a print by Gaywood

IVORY MORTAR AND PESTLE CARVED IN RELIEF WITH
A REPRESENTATION OF A LABORATORY
The retort is connected with a water-cooled condenser (see p. 115)
Italian, seventeenth century

ALCHEMISTS OF THE XVITH CENTURY

the mines of Sigismund Fugger, he acquired a knowledge of metals and ores and studied the diseases of his fellow-workers. He was gifted by nature with original talents, and saw from his experience that the practice of medicine had degenerated to a mere pretension and relied chiefly on purging, bleeding, and emetics. He realized that Nature was the best healer and that the best results could be achieved by natural methods. His fame as a physician spread rapidly, and in 1527 he was appointed Professor of Medicine at the University of Basle.

Although he became popular with his students he made many enemies, chiefly on account of his derision of the doctrines which had been practised since the time of Hippocrates. The burning of the books of the Fathers of Medicine in the presence of his students was typical of the man who declared that he had more knowledge in his bald pate than was in all their writings, and that in the buckles of his shoes there was more learning than in Galen and Avicenna, and in his beard more experience than in all their universities. Though he was undoubtedly conceited and egotistic, he was loved and respected by his pupils, who called him their "dear preceptor and king of arts." Driven at length out of Basle by the jealousy of his rivals, he turned his attention to alchemy and sought to study the art under Trithemius, the Abbot of Spannheim. It was from about this time that he took the name of Paracelsus ('Greater than Celsus'[1]), or, as he later styled himself, Theophrastus Paracelsus, "Prince of Philosophy and Medicine." It is his work in alchemy which chiefly concerns us, for through his labours the school of iatro-chemistry was founded, and new paths for chemistry and medicine were opened up which gave an impulse to the study of medical chemistry that continued down to the beginning of the nineteenth century.

Iatro-chemistry had for its aim the investigation of chemicals used in medicine and the investigation of their composition and

---

[1] Aulus Cornelius Celsus was a Roman of the first century A.D. whose writings on medicine and surgery are still held in honour.

action. Paracelsus taught that the object of alchemy was not to make gold, but to prepare medicines. The alchemist was to discover the medicines and prepare them, while the physician was to examine and explain their action. He believed that the human body was a combination of certain chemical matters; should these undergo changes, diseases resulted, to cure which

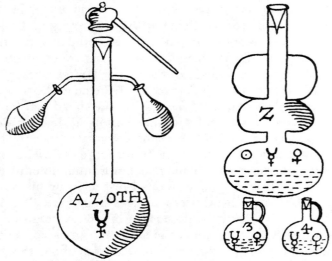

APPARATUS FOR TRANSMUTATION DESCRIBED BY PARACELSUS
From woodcuts of the sixteenth century

chemical medicines were necessary. He thus discarded all the principles that had been for centuries the basis of the art of medicine.

He attributed disease to a disproportion in the body between the quantities of the three great principles—sulphur, mercury, and salt—which he regarded as constituting all things. He considered that an excess of sulphur was the cause of fever, and contended that if illnesses were caused by chemical changes in the body they could only be cured by chemical substances.

He thus threw over all the old theories that had influenced

ALCHEMISTS OF THE XVItH CENTURY

the art of medicine from the time of Galen. He set out to fight against the ancient doctrines of the Fathers of Medicine with courage and vigour, but his views were received with opprobrium, and he was deemed by the orthodox professors a "bombastic charlatan." In replying to some of his opponents he thus compared the alchemists with the physicians. The former, he says,

> are not given to idleness, nor do they go about in a proud habit or plush and velvet garments, often showing their rings upon their fingers or wearing swords with silver hilts by their sides or fine and gay gloves upon their hands, but diligently follow their labours, sweating whole days and nights by their furnaces. They do not spend their time abroad for recreation, but take delight in their laboratories. They wear leather garments with a pouch and an apron wherewith to wipe their hands. They put their fingers amongst the coals into clay and filth, not into gold rings. They are sooty and black, like smiths and colliers, and do not pride themselves upon clean and beautiful faces.

Paracelsus had the courage of his convictions and boldly prescribed arsenic, antimony, and mercury in the treatment of certain diseases. He used almond-oil as a solvent for the essential oils he extracted from certain drugs, and knew the value of alcohol as a solvent for the properties of vegetable bodies. "There is no better way," he states, "of extracting the essence of roots and herbs than to cut them up as small as possible and boil them in strong wine in a closed vessel, separate them by straining and distil the liquid through an alembic."

Judging from his works, he appears to have been familiar with nearly every chemical preparation known in his time; he held that chemistry was indispensable to medicine and an essential part of medical education. He laboured for years to discover some method of prolonging human life, for he fully believed that the human body could be rejuvenated. "Metals may be preserved from rust," he observes,

> wood may be protected from rot, blood may be preserved a long time if the air is excluded; Egyptian mummies have kept their

ALCHEMY AND ALCHEMISTS

form for centuries without undergoing putrefaction; animals awaken from their winter sleep, and flies, having become torpid from cold, become nimble again when they are warmed; therefore if inanimate objects can be kept from destruction, why should there be no possibility to preserve the life-essence of animate forms?

He foreshadowed the discovery of alkaloids and other active principles of plants. "There is," he remarks, "a force of virtue shut up within things, a spirit like the spirit of life, in medicine called Quintessence or the spirit of the thing." He also believed that there were certain specifics in the treatment of disease. Thus his specific for causing sleep consisted of opium with orange and lemon-juice flavoured with cinnamon and cloves. He distinguished between the purgative action of vegetable drugs like rhubarb and colocynth and the salines such as potassium and sodium sulphate. His purgative specific was chiefly composed of colocynth, and his diaphoretic contained camphor, cardamoms, pepper, and grains of paradise, from which it will be noticed that the active drug in each of these preparations had an undoubted specific action.

"If good sometimes turns to evil, then it is possible to make evil things good," he observes in one of his treatises.

> A substance may be poisonous, yet so employed that it will not act as a poison. For example, arsenic is the most poisonous of substances and a drachm of it will kill a horse, but fire it with salt of nitre and it is no longer a poison; a horse may then swallow ten pounds without harm.

He claimed that the function of medicine was to supplement Nature, and said, "God makes the true physician, but not without pains on man's part."

In spite of his critics and traducers, Paracelsus was a man of genius and a real seeker after truth; a careful study of the voluminous works attributed to him compels the student to form a high estimate of his great knowledge and character.

ALCHEMISTS OF THE XVIth CENTURY

Like many other reformers, he ended his days in poverty. He drifted from place to place, and at length, on the invitation of the Archbishop, he went to Salzburg, where he died at the age of forty-eight. He is buried in the cloisters of the ancient church of St Sebastian in that city, and a stone set in the wall bears the following inscription:

PHILIPPI THEOPHRASTI PARACELSI
QUI TANTUM ORBIS FAMAM EX AURO CHYMICO ADEPTUS EST EFFIGIES ET OSSA DONEC RURSUS CIRCUMDABITUR PELLE SUA.

Job ch. xix

SUB REPARATIONE ECCLESIA
MDCCLXXII
EX SEPULCHRALI TABE ERUTA HEIC LOCATA SUNT.
CONDITUR HIC PHILIPPUS THEOPHRASTUS INSIGNIS MEDICINE DOCTOR. QUI DIRA ILLA VULNERA. LEPRAM PODAGRAM HYDROPOSIM ALIAQ INSANABILIA CORPORIS CONTAGIA. MIRIFICA ARTE SUSTULIT. AC BONA SUA IN PAUPERES DISTRIBUENDA COLLOCANDAQ HONERAVIT. ANNO M.D XXXXI DIE XXIIII. SEPTEMBRIS VITAM CUM MORTE MUTAVIT.

PAX VIVIS REQUIES
AETERNA SEPULTIS.

Paracelsus' ring and autograph are preserved with other relics in the museum of the beautiful city in which he ended his days. His fame as a healer still survives in the district, and on the anniversary of his death sick people still make pilgrimages to his tomb in the hope of obtaining relief from their sufferings.

REPUTED AUTOGRAPH OF PARACELSUS
From a manuscript preserved at Salzburg

The works of Johann Isaac Hollandus and his son, printed in 1572 and subsequent years, appear to have been largely taken from the treatises attributed to Paracelsus. Nothing is known of their lives, and it is considered that their works, if authentic, were written after his time. Boerhaave says they were natives of Stolk, and Schmieder gives reasons for believing that they lived in the early fifteenth century. The writings which pass under their authorship are commended by Boerhaave, who says that Isaac was skilful in enamelling and in imitating precious stones.

In considering the advance of alchemy in the sixteenth century the work of Georg Agricola, or Bauer, who was born at Glauchau, in Saxony, in 1490, should not be forgotten. He devoted his energies mostly to metallurgy, and describes his researches in a work entitled *De Re Metallica*, printed in 1556. He describes bismuth, distinguishing it from tin and lead, and gives a clear account of the amalgamation process for the extraction of gold. He indicates the possibility of flame-tests by noticing the characteristic colours of certain metals when heated in a flame.

Another alchemist who contributed to the knowledge of the time was Andreas Libavius, who was born at Halle in 1540. He first studied medicine and practised as a physician, but, having accepted the iatro-chemical theory, turned his attention

ALCHEMISTS OF THE XVItH CENTURY

to the discoveries of mineral bodies that might be employed in medicine. His work entitled *Alchymia dispersis passim optimorum auctorum veterum et recentium exemplis potissimum* is regarded by some as the first text-book on chemistry. He was the first to make the chloride of tin still known as *spiritus fumans Libavii*, and his name is thus perpetuated.

CHAPTER XIX

CÆTANO AND BORRI

IN the early part of the seventeenth century the ranks of the pseudo-alchemists were increased by a number of charlatans and clever rogues, who attained notoriety by their impudent pretensions to transmute the baser metals into gold and travelled the countries of Europe in search of wealthy patrons. One of the most extraordinary of these adventurous characters was Domenico Manuel Cætano, the son of a mason of Petrabianca, near Naples. He was apprenticed to a goldsmith in Naples, where he no doubt acquired some knowledge of the precious metals. He also learned to become an expert conjurer.

Of a roving disposition, Cætano bade farewell to his master and set out to travel. After traversing Italy he went on to Spain, and for four months remained in Madrid, where he began to practise as an alchemist. During this period he contrived to make several influential friends, among whom were the Bavarian Envoy at the Spanish Court and his cousin, from whom he obtained a considerable sum of money. He persuaded the Envoy to give him letters of recommendation to important people in Germany, describing his skill as an alchemist, and, thus equipped with introductions and money, he set out for Bavaria. On arrival he presented himself at the Court of the Elector Maximilian Emmanuel, who was then Viceroy of the Netherlands, and with the Envoy's recommendations soon obtained an audience. He told the Elector that he knew the secret of transmutation whereby he could convert the baser metals into gold or silver, and was ready to make considerable quantities if he were supplied with sufficient money to prepare the 'tinctures.'

The Elector, anxious to retain the services of so promising

CÆTANO

an adept, at length consented to give him 60,000 florins and appointed him to several posts at the Court. Cætano expressed his delight and started to equip a suitable laboratory, but months went by and there were no results. At length the Elector began to get impatient of the long delay and demanded that Cætano should give a demonstration of his skill, but each time he was asked to fix a date he had always an excuse ready and no gold was forthcoming. Finding himself under suspicion, he resolved to attempt to leave the city, but he was caught when trying to escape, and on being convicted of deception he was imprisoned in the castle of Grünewald. After being kept in close confinement for six years he managed to effect his escape, and for a while disappeared.

Some time afterward a mysterious Count de Ruggiero arrived in Vienna and established himself in a fine suite of apartments. It soon became known that he was an expert in alchemy and by this means had acquired great wealth. He rapidly made friends, and at length obtained an introduction to Prince Anthony of Liechtenstein, whom he succeeded in convincing that he knew the secret of changing the baser metals into gold. The Emperor Leopold, on hearing this, decided to take the Count into his service and offered him a large salary. This was just what Ruggiero, otherwise Cætano, wanted, and on receiving an audience with the Emperor he boldly asked him for a *grant* of a sum of money so that he could prepare his tinctures, but before he received it the Emperor died and Ruggiero's pay was stopped. However, he managed to ingratiate himself with the Dowager Empress, who gave him 6000 florins, together with a recommendation to John William, the Elector of the Palatinate. He told the Elector that if he did not produce 72,000,000 florins by means of his tinctures within six weeks he would willingly forfeit his head. John William agreed to give him the time he asked, and Ruggiero was supposed to have set to work, but instead of making his tinctures, he appears to have spent his time in making love to the pretty daughter of a Viennese midwife. The end of the six weeks drew near, and, as he had no desire to lose his head, he made his plans and fled in the night, accompanied by the lady.

It is not until a year afterward that we hear of the Count again, when, as Cætano, he made his appearance in Berlin. He had obtained money by some means and rented a large house in a fashionable part of the city, driving out in a magnificently gilded carriage drawn by four horses. This gorgeous equipage and the style in which he lived aroused great interest, and the society of the wealthy Count was soon sought by many of the prominent citizens.

Having made a favourable impression, he next proceeded to present a petition to King Frederick in which he implored his protection against the persecution of foreign Powers to which he was then being subjected. In return he offered to enrich the Royal Treasury with untold gold, which he alone could make by his secret process, and was willing to demonstrate his powers if the King desired him to do so.

Although King Frederick, like other monarchs of the time, was not averse from having his coffers so easily replenished, he decided to proceed with caution, and requested Deppel, a Danish alchemist of repute who was then living in Berlin, to find out what he could about the Count. Deppel soon became on friendly terms with Cætano, and one day persuaded him to produce some of his precious "red and white tinctures," of which he had a small quantity. Deppel was apparently sceptical at first, but Cætano managed to deceive him into the belief that he could transmute mercury into silver.

The King, favourably impressed by Deppel's report, sent Cætano a message commanding him to make a demonstration of his powers in the royal presence. To this Cætano at once agreed, and a day was fixed for the King's visit to his laboratory, where the operation was to be carried out. Cætano cunningly laid his plans, as he knew that the result would mean either fortune or an urgent need for flight.

CÆTANO

At the time appointed the King, accompanied by the Crown Prince, arrived at Cætano's house, and he conducted them with much ceremony to his laboratory. The royal visitors, in rich costumes of velvet and silk, were attended by their gentlemen-in-waiting and accompanied by some goldsmiths who had been brought to test the metal. An attendant then brought in a vessel containing a quantity of mercury which the Crown Prince had undertaken to provide. When all was ready the King seated himself where he could watch every movement of the operator, and the others gathered round in eager expectancy.

Going to a large crucible which stood upon a hearth, Cætano slowly poured some of the mercury into it and, placing it on a sand-bath, proceeded to blow up the fire. He heated it for some time; then, taking a small phial which contained a thick, reddish-coloured liquid, he added a few drops of this to the mercury, meanwhile carefully stirring the contents of the crucible, from which dense yellow fumes soon began to rise. After continuing the heat for a space of half an hour Cætano, taking his tongs, lifted the crucible from the fire and placed it on a slab to cool. The minutes passed in silence; then, after a while, he invited the visitors to look into the crucible. The King and the Crown Prince stepped forward and, glancing at the contents, were astonished to see a mass of metal which Cætano declared was gold. The crucible was then broken and the metal handed to the incredulous goldsmiths, who, after applying their tests, declared it to be real gold.

Cætano is said to have then carried out two other operations, in one of which he changed a quantity of mercury into silver by means of his white tincture and in the other with his red tincture transmuted part of a bar of copper into gold.

The King congratulated Cætano on the success of his experiments and expressed his satisfaction at the results; whereupon Cætano begged him to accept a small quantity of his tinctures, and promised to hand over to him within sixty days eight ounces of the red and seven ounces of the white, by means of which he declared gold and silver to the value of seven million thalers could be produced.

As soon as the result of the King's visit was known Cætano became the most popular person in Berlin. People of all classes flocked to his house, and invitations to banquets and entertainments were showered upon him. But Cætano refused them all on the pretext that he was busy preparing the tinctures that he had promised to deliver to the King within two months, and denied all visitors. Shortly before the time expired he secretly left Berlin and went to Hildesheim, whence he wrote to the King stating that he was ready to impart his secret to anyone whom his Majesty might think proper to select.

The King, astonished to find from the letter that Cætano had left the city, at once sent his Chamberlain, von Marschall, to Hildesheim to obtain the secret and to bestow a present of his miniature set in diamonds. At the same time he was instructed to hand the Count a commission appointing him a Major-General of Artillery. A battle of wits then began between the Chamberlain and Cætano, who declared that he would not hand over his secret until he received a thousand ducats in payment, and in the end the Chamberlain decided to return to Berlin and lay his report before the King.

After he had departed Cætano hurriedly left for Stettin, and on his arrival there wrote to the King stating that the Herr von Marschall had treated him badly. After learning his secret he wished to keep it for himself, and that he was an unfaithful servant. He asked that the thousand ducats should be sent to him without delay.

On receipt of this letter the King despatched his Privy Secretary, Hesse, with instructions to pay him a sum of money and to try to induce him to return to Berlin. But the efforts of Hesse were unavailing, and directly he had left Stettin Cætano set off for Hamburg.

In the meantime the King had received information from Vienna that the Bavarian Envoy had declared that Cætano had

CÆTANO

swindled his cousin in Madrid out of a large sum of money. On learning this he determined to examine the phials of tinctures that Cætano had given him, and when they were brought before him they were found to be empty!

Enraged at this discovery, the King gave orders for the immediate arrest of Cætano on the charge of disobeying his orders as a Major-General. The trickster was arrested in Hamburg and brought as a prisoner to Berlin. He begged that he might be allowed to continue his experiments in the presence of a commissary, and undertook to prove that he could carry out all he claimed to do. His petition was granted, and means were given to him to carry on his work under strict surveillance. To the amazement of the commissary who was in charge, one day he produced a quantity of silver which he declared he had made by means of his white tincture. This was shown to the King, who was so delighted that Cætano was at once restored to favour and apartments were provided for him in the palace, where a watch could be kept on him so that he could not escape. He was furnished with fine clothes, and the Court cook was ordered to serve him with "ten special dishes for dinner and eight more for supper."

But the King was determined not to let his 'gold-maker' remain idle now that he was supplied with every necessity, and at length obtained a solemn promise from him that he would transmute a hundredweight of mercury into gold by a given date.

Although Cætano undertook, with his customary boastfulness, to carry out the task, he became panic-stricken as the time approached, and when the day arrived he could not be found. By some means he had managed to elude his guard and escape from the palace. A search for the missing alchemist was immediately instituted, but no trace of him could be found, and it was not until some time afterward that he was discovered to be living in Frankfort-on-Main. Here, at the request of the Prussian Minister, he was once more arrested and lodged in prison. There he remained for some time until he was eventually handed over to an escort of Prussian troops and taken to Cüstrin, where he was brought to trial on the charge of treason, was found guilty, and condemned to death. He was hanged in the market-place on August 29, 1709, the beam of the gallows in grim irony being gilded with Dutch metal; and when his body was cut down it was paraded through the streets dressed in a golden robe.

So ended the strange and adventurous life of Domenico Manuel Cætano. He was a charlatan and impostor from the beginning, and yet as late as the last century there were many who still believed that his claim to possess the secret of transmutation was genuine and that he fell a victim to his own boasting. Cunning though he undoubtedly was, his effrontery was amazing, and he appears to have deceived his patrons with the same story time after time. His career is instructive as showing the greed for gold and the credulity that existed among personages in high places in Europe late in the seventeenth century.

A man of a different type but one who had a most eventful life was Giuseppe Francesco Borri, who was born at Milan in 1627. His father was a physician, and sent his son to be educated at the Jesuits' College at Rome. Here he soon showed remarkable talent and had no difficulty in acquiring a knowledge of all branches of learning, but was especially attracted to the study of medicine and alchemy. Continuing his medical studies after he left the college, he is said to have led a very loose life for a time, but on settling down he began to practise medicine in Rome and eventually obtained a position in the Pope's household.

When he was about thirty-seven he again took up the study of alchemy, and in 1653 became private secretary to the Marquis di Mirogli at Rome. From being a freethinker, he developed a strong religious tendency to the extent of fanaticism, and expressed a belief that the secrets of the Omnipotent and of Nature had been revealed to him. He began to deliver lectures questioning the supremacy of the Pope, and claimed that the mysteries of faith were derived from the principles of alchemy. He soon had to quit Rome, and first went to Innsbrück, leaving there for Milan with the intention of establishing a new religion. Here he frequently preached his doctrines and found many followers, whom he told that he had received from the Archangel Michael "a heavenly sword upon the hilt of which were engraved the names of the seven celestial intelligences."

In their enthusiasm he and his followers even attempted to take possession of Milan, but their plot was discovered and Borri escaped to Switzerland. An order was obtained for his arrest through the Inquisition, and a reward of 35,000 francs was offered to anyone who would deliver him up. Meanwhile he was tried in his absence and condemned to death as a heretic and sorcerer, his effigy being burned in Rome by the common hangman in 1661.

For a time Borri lived at Strasburg, and then journeyed north to Amsterdam, where he established himself in a fine house and assumed the title of 'Excellency.' Here he practised medicine with considerable success and visited his wealthy patients with great pomp, riding in a gilded coach. Meanwhile he continued to carry on operations as an alchemist. In the latter capacity he succeeded in obtaining 200,000 florins from a rich merchant on the pretext that he was on the verge of discovering the Elixir of Life. As this did not mature and he could not repay the loan, he suddenly left one night for Hamburg, where at that time Christina, the ex-Queen of Sweden, was living. He knew that she was interested in alchemy, and he hoped to obtain her patronage in pursuing his quest of the Philosopher's Stone, but, being warned that Hamburg was unsafe for him, he journeyed on to Copenhagen to seek the protection of Frederick III, King of Denmark, who was also a believer in the art.

The King, being in want of money, agreed to provide Borri with the means to carry on his experiments and took a great interest in his plan of operations. Borri set to work, and though delay after delay occurred in his efforts to fulfil his promise to produce gold, he managed to ingratiate himself with the King, and by making himself useful to the monarch in other ways he gained his good opinion. He thus spent six years at the Danish Court, and on the death of the King in 1670 he set off to travel again. He first went to Saxony, but, fearing the emissaries of the Inquisition, he decided to proceed to Constantinople, where he thought he would be out of reach of the Papal authorities. Just at this time Leopold I, Emperor of Austria, was suffering from a mysterious illness which greatly puzzled his physicians. According to a story related by Wraxall, the historian, one day, while the monarch was in consultation with the Papal Nuncio concerning an insurrection which had broken out in Hungary, a despatch arrived containing a list of the persons implicated, and among them appeared the name of Francesco Borri. As the name was read out by the secretary the Nuncio started and exclaimed, "Borri! Have him arrested at once, your Majesty. He is a most dangerous man and has contrived to escape from the avenging arm of the Holy Office." Within a few hours a Captain Scotti was despatched on a special mission to Goldingen to arrest him.

As it happened, Borri had arrived at Goldingen on the Silesian frontier on April 10 and was compelled to reveal his real name when, being suspected of being connected with the conspiracy, he was arrested. Thus his name was included in the list of suspects sent to Vienna. On Captain Scotti's arrival Borri was handed over to him as a prisoner and, travelling in a carriage with an escort of cavalry, the party at once set out for Vienna. Scotti, being an Italian, treated his captive with every consideration, and on the journey told Borri that he was suspected of being concerned in the conspiracy and that he had the Papal Nuncio among his opponents. "Then I realize the real cause of my arrest," said Borri.

Scotti also told him of the Emperor's mysterious illness and remarked it was now supposed to be due to secret poisoning.

BORRI

Borri declared that if this was the case he could readily discover the presence of a poison, should one exist, and implored Scotti to inform the Emperor that if he really suspected poisoning he could find the cause. Scotti promised to comply with his request.

On their arrival in Vienna on April 28, Borri was taken to the Swan Inn and lodged in a room which was guarded by soldiers. Tired and wearied by the journey, he at once threw himself on the bed and fell asleep, but he was aroused during the night by the door being opened. A man wrapped in a cloak and carrying a lantern entered, and this midnight visitor Borri recognized as Captain Scotti.

"Make haste and get ready," said the Captain in a low voice. "The Emperor wishes to see you, for your reputation as a physician is known to him. His Majesty trusts you, but I was compelled to wait till night as he does not wish this visit to be known."

In a few minutes the two men were walking through the dark and silent streets toward the palace. When they arrived Scotti handed his prisoner over to a chamberlain, who conducted him to the Imperial antechamber and bade him be seated. In a few minutes a gentleman of the bedchamber entered and made a sign to Borri to follow him. They passed through several apartments until they came to a velvet-covered door, which the gentleman opened, and, drawing back a heavy *portière*, he beckoned Borri to enter.

The Emperor's cabinet was a gloomy room lighted by a few candles, which shed but a dim glow on the pictures which covered the walls. Seated in an armchair near the table, a little man was discernible wrapped in a dressing-gown of green silk and wearing a cap with a shade for his eyes. His feet, with which he was making impatient movements, rested on a stool; his face was livid and his cheeks shrunken.

Borri took a step forward and bowed, and the little man looked up.

"Are you Francesco Borri?" he asked, in a trembling voice.

"At your Majesty's service," replied Borri.

"I am sorry to see you here as a prisoner, but you are not one at present," said the Emperor.

"Had I not been arrested I should not have had the happiness of seeing your Majesty," rejoined Borri.

"I hear much that is satisfactory about your learning, although in another respect you are said to be a dangerous man. Why do you trouble yourself with religious affairs? Leave them to the clergy. I hear that you now devote yourself to medicine. What have you heard about my condition?"

"Nothing beyond the supposition that your Majesty is being poisoned," replied Borri. "But, that I may be able to express my views on the subject, your Majesty's physician must tell me of your symptoms, and then I shall be able to speak with certainty."

A messenger was at once sent for the physician. Borri meanwhile was struck with the Emperor's grey and wasted appearance, and, rising from his chair, he took a survey of the room, examining every ornament and object and smelling them with suspicion.

The Emperor followed his movements with inquiring eyes.

"Well, Borri," he sighed at length, "what do you think?"

"I think that almost certainly your Majesty is being poisoned," said Borri decisively.

"Holy Mother, have mercy on me!" cried the Emperor.

"I must first speak with your physician, but I think I can promise your Majesty's recovery with equal certainty, for there is still time."

"How do you come to this conclusion of poison? My friends dine with me and eat the same food. Do you notice anything on my body?"

"It is not so much your Majesty's body as the atmosphere of your room that is poisoned," observed Borri.

"How can you tell, when I feel nothing of it?"

BORRI

"Your Majesty is too accustomed to the poisonous exhalation to notice it."

"And whence comes the exhalation?" asked the Emperor.

Borri rose and, taking the candelabra that lighted the room, placed them all on the table.

"See the exhalation that rises from the candles!" he exclaimed. "Do you not notice the peculiar colour of the flame?"

At this moment the chamberlain entered the room, and the Emperor asked him if he noticed the smoke arising from the candles, and he replied that he did.

The physician then arrived. "You have come at the right moment," said the Emperor. "It is asserted that the air of my room is poisoned. Give me the report of my illness."

The Emperor passed the document to Borri, who glanced quickly at it and nodded his head.

"Do you not perceive the curious smell in the room, and the fine, quickly ascending vapour from the burning candles?" Borri asked the doctor. "It would be interesting to know if the same candles are used in the Empress's apartments."

The chamberlain at once brought two lighted candles from the Empress's chamber and placed them on the table beside the suspected ones. The former burned clear and quietly, while the latter had a ruddy flame and emitted a thin vapour, and occasionally sparks flashed from the wicks.

"There is the cause of your sickness!" exclaimed Borri, as he laid his hand on a candelabrum. "And I will prove to your Majesty that these are impregnated with a subtle poison."

Extinguishing the suspected candles, Borri removed all the wax from the wicks and, shredding the wick of one, asked that it should be mixed with some meat and given to a dog. The turnspit dog was brought and shut up in a cupboard with the dish of meat. Meanwhile the Emperor was removed to another apartment, and Borri and the physician proceeded to the palace pharmacy to prepare an antidote for him. Here also Borri tested the suspected candle-wick and found, as he had thought, that it was impregnated with arsenic. He had left instructions that he was to be called as soon as the dog got restless, but the animal was found to be dead by the time he returned to the Emperor's cabinet.

The antidote prepared by Borri soon produced a beneficial effect on the Emperor, and his health improved so rapidly that within three weeks he was able to go out again.

The record of Borri's examination of the candles is interesting and shows that he was a sound chemist. He examined the whole of the suspected materials and weighed them. The weight of the candles was twenty-four pounds, and of the wicks three and a half pounds, from which, by comparison with uncontaminated candles, he estimated that nearly two and three-quarter pounds of arsenic had been employed to impregnate them.

For a short time the Emperor appears to have treated Borri well, and he dined at the Imperial table; but the hatred of the clerical party toward him was increased when they saw him thus favoured. On June 14, 1670, the Emperor, now quite restored to health, summoned Borri to his cabinet and thanked him fervently for his services, but added he was sorry that in the matter of religion he had gone astray. "The Pope will appoint a commission," he continued, "and I have obtained a guarantee from the Papal Nuncio that in no case shall anything be done against your body or your life." Further he promised Borri a pension of two hundred ducats a year as an award for his services. He was then handed over to the clerical authorities, and on the following day was sent under an escort to Rome.

He was lodged in the prison of the Inquisition until he consented to make a public recantation of his heresies, which he did before great crowds of people on October 27, 1672. This act saved his life, but he was condemned to perpetual imprisonment.

It is stated that, through the intervention of the Duc d'Estrées, whom he had cured of a disease which baffled all his

BORRI

physicians, he was transferred to the Castle of St Angelo, where he was allowed more freedom and also permitted to pursue his studies in alchemy. Here he remained in a cell for twenty-three years, carrying on his work, and writing a book dealing with the Rosicrucian philosophy which was printed in Cologne. Through the influence of Queen Christina, who was allowed to visit him, he received considerable indulgence and was permitted to have apparatus in his cell so that he could carry on his experiments. She also provided him with money and encouraged him to continue his researches in the hope that he would at last find the "Great Secret."

Borri lived to the age of eighty and died in the Castle of St Angelo, where the cell in which he lived and carried on his work is still shown.

CHAPTER XX
ALCHEMISTS OF THE SEVENTEENTH CENTURY

CONSIDERABLE mystery surrounds the life of Alexander Seton, a reputed Scottish alchemist, and many of the stories told concerning his alchemical exploits are doubtless fictional. He is first heard of in 1601, when, it is said, he was living at Seton Hall, near Edinburgh, on the east coast of Scotland.

One stormy night a Dutch vessel was wrecked on the beach not far from the hall, and the captain and some of the crew were saved by the help of Seton, who took them to his house near the seashore and treated them with great kindness. After they had recovered he supplied them with means to return to Holland. One of the men, named James Haussen, who is said to have been the pilot of the ship, in gratitude begged Seton to come and see him at Enkhuysen, where he lived. The following year Seton sailed to the Netherlands and found Haussen's abode, where he was received with great joy and entertained for several weeks. The bond between the two men appears to have been the study of alchemy, as it is related that during the visit Haussen was the astonished witness of several transmutations which were performed by his guest, who confessed that he was an adept in alchemy.

Before the end of the visit Seton revealed to his host the secret of transmutation, and in his presence converted a piece of lead into gold of the same weight. After he had left Haussen confided his experience to a certain physician in Enkhuysen, and gave him a piece of the gold which Seton had produced in his presence on March 13, 1602. This gold eventually passed into the hands of the doctor's grandson.

ALCHEMISTS OF THE XVIIth CENTURY

On leaving Enkhuysen Seton went to Amsterdam and Rotterdam and thence embarked for Italy. He afterward journeyed through Switzerland, where he met Dr Wolfgang Dienheim, a professor of the University of Fribourg, and Dr Jacob Zwinger. Dienheim describes Seton as "a short man of very spiritual appearance, very stout, his face of a high colour" and as "wearing a beard in the French style." Although Dienheim had at first no belief in the doctrine of transmutation, Seton is said to have convinced him of its truth by ocular demonstration.

This demonstration, which took place in the workshop of a goldsmith at Basle, was carried out in the presence of several of his friends. The professor, in his description of the operation, says:

> They took with them some sheets of lead, which were bought by Jacob Zwinger, together with some sulphur, on his way from his house, and a crucible was purchased from the goldsmith. In the workshop Seton handled nothing, but made a fire in the furnace and melted the lead and sulphur together in the crucible and stirred them with an iron rod.

In a short time Seton asked the professor to throw on the molten metal in the crucible a heavy yellow powder which he had in a piece of paper. Dienheim says:

> Though unbelieving as St Thomas we did as directed, and in fifteen minutes the crucible was removed from the fire; on cooling, the lead had disappeared and a button of gold remained which the goldsmith pronounced to be superior to that of Hungary or Arabia.

It weighed as much as the lead, and the two doctors were amazed but convinced, as Seton had done nothing himself beyond supplying the small packet which contained the yellow powder of projection. Seton had a piece of the gold cut off weighing four ducats and gave it to Zwinger to keep.

On leaving Basle Seton travelled under the assumed name of Hirschborgen to Strasburg, where he found lodgings with a merchant called Koch and performed a transmutation in his presence. He soon left for Cologne, where he stayed with Anton Bordemann, a German alchemist of that city. Here he carried out further successful operations and then went on to Munich, where he met and fell in love with a Bavarian girl of great beauty, whom he married.

The stories of his success in making gold attracted the attention of Christian II, the Elector of Saxony, who summoned Seton to his Court. For some reason the alchemist did not answer the invitation in person, but sent his assistant, Hamilton, in his place. The latter is said to have performed a successful projection in the presence of the whole Court, the gold he made standing every test.

The Elector was now eager to see Seton in person and sent him an urgent command to appear at the Court. Assuming the name of "the Cosmopolitan," he set out for Dresden, where he was received with honour and distinction. He presented the Elector with a small quantity of his red tincture, but refused to divulge the secret of his method of transmutation. Attempts to persuade him to reveal his process proved fruitless, and at length the Elector, incensed by his obstinacy, threatened him with torture. This again was in vain, so he was arrested and imprisoned in a tower near the city. Here he was guarded by forty soldiers, and after being submitted to the torture was thrown into a dark cell and left in solitary confinement.

It so happened that at this time Michael Sendivogius, a Polish alchemist, was paying a visit to Dresden, and, hearing of Seton's plight and suffering, he obtained permission to see him. Later, by bribing his guards and after obtaining the promise of Seton to help him in his work, he contrived his escape. They fled together to Cracow, where Sendivogius had some property, but on their safe arrival there Seton refused to part with his secret, as his rescuer had hoped he would do, giving the reason that "the revelation of such an awful mystery would be a heinous sin."

He died some two years later, but all he left to his preserver

ALCHEMISTS OF THE XVIIth CENTURY

were the remains of his coveted red tincture, and his secret was never revealed. Sendivogius married Seton's widow, who possessed two manuscripts said to have been written by her first husband, one being a treatise entitled *A New Light on Alchemy*, which Sendivogius afterward published as his own work, and the other being the *Twelve Treatises of the Cosmopolitan*.

Sendivogius then set out on his travels with the small amount of Seton's 'tincture,' or 'powder,' of projection which he possessed, and is said to have made several successful transmutations in public in different cities. He soon became famous, and all the rulers of the countries of Central Europe sought a visit from the celebrated alchemist. Among them was the Emperor Rudolph, before whom, we are told, Sendivogius made a successful transmutation. The alchemist gave him a small quantity of the tincture, and the Emperor is said to have carried out a successful experiment with his own hands. To commemorate this achievement he had placed on the wall of the room a marble tablet, which was thus inscribed:

> Who'er could do under the rolling Sun
> What Sendivogius the Pole hath done?

Sendivogius was made a Counsellor of State, and a gold medal was struck in his honour. When at length he obtained permission to leave Prague he set out for Cracow, but, according to one account, he was seized on the road by a Moravian noble, who made a prisoner of him with the object of robbing him of his precious tincture. With the aid of a file, however, he cut the bars of his prison window, and, by tearing up some of his clothing, made a rope, by means of which he escaped. He appealed to the Emperor, who confiscated the nobleman's estate and gave it to Sendivogius, who lived on it for some years.

Bodowski states that Sendivogius kept his wonderful tincture in "a little box of gold, and when on a journey hung it round his neck, but the greater part was kept secreted in a hole cut in the step of his carriage."

Sendivogius at length returned to Warsaw and remained there until he received an invitation to visit the Court of Würtemberg. On arriving at Stuttgart, however, he met with a rival in Johann Heinrich Müller, who had also figured at Rudolph's Court at Prague, but had since entered the service of the Duke of Würtemberg. Fearing that he might be displaced by the Pole, Müller gathered a band of armed horsemen, and they waylaid Sendivogius on the road, arresting him in the name of the Duke. They stripped him of his clothing, bound him naked to a tree, and made off with his gold box containing the tincture, as well as a diamond-studded cup and his gold medal. The unfortunate alchemist, who was released by some passing travellers, made a complaint to the Emperor, who demanded of the Duke the person of Müller and the restoration of the stolen property. The Duke, becoming alarmed, had Müller hanged in the courtyard of the palace in 1607 and restored the cup and medal to their owner; but the gold box was never found.

According to another historian, Sendivogius had kept a very small quantity of the tincture carefully concealed, and on returning to Cracow dissolved all that was left of it in some rectified spirit of wine. With this solution he astonished the physicians of the city by making some amazing cures—among them of Sigismund III, King of Poland, who was suffering from the effects of a serious accident. All his media now being exhausted, Sendivogius is said to have become a wandering charlatan, eventually dying in 1646 at the age of eighty-four.

Among other mysterious adventurers of this period was a person named Lascaris, who represented himself as archimandrite of a convent in the island of Mitylene. At the age of about forty he is described as "of attractive mien, with agreeable manners and fluent in conversation." He is first heard of in Berlin, where he was taken ill and sent for an apothecary. The latter was unable to visit him, but sent his apprentice, a young man named Johann Friedrich Bötticher. Lascaris took a great liking to the young apothecary, and, in gratitude, on

SYMBOLIC FIGURE REPRESENTING AN ALCHEMIST AND HIS WIFE ENGAGED IN THE PREPARATION OF THE PHILOSOPHER'S STONE

Mutus Liber (1677)

AN ALCHEMIST AND HIS WIFE ENGAGED IN
VARIOUS OPERATIONS

Mutus Liber (1677)

ALCHEMISTS OF THE XVIIth CENTURY

his recovery gave him a small quantity of the powder of projection, telling him that he was not to mention from whom it had been derived, and on no account to use it until after his departure.

With this valuable possession, Bötticher resolved to give up medicine and devote himself to alchemy. He convinced his friends of his wisdom by transmuting some silver into gold in their presence, and he soon became famous in Berlin. Frederick I, on hearing of his operations, summoned him to his presence, but Bötticher, knowing how badly many alchemists had been treated by their royal patrons, fled the city and took refuge with an uncle at Wittenburg.

The Elector of Saxony was the next who desired to see him, and Bötticher decided to proceed to his Court. He was received with great favour, and was successful in convincing the Elector that he could convert base metals into gold. The title of Baron was conferred upon him, and he took a fine house in Dresden and lived in great style. All went well for a time until his supply of the powder of projection, the source of his wealth and fame, was running low, and, what was worse, he did not know how to replenish it. By the time his last grain was expended he was heavily in debt, and was planning to leave the city when the Elector, hearing of it, had his house surrounded by soldiers and held him a prisoner.

The mysterious Lascaris now emerged from his seclusion, and, hearing that his *protégé* Bötticher was in trouble, he employed a young physician named Pasch to try to liberate him. Pasch, in the course of his negotiations for the release of Bötticher, was also made a prisoner, and was confined in the fortress of Sonnenstein, while Bötticher was imprisoned in the castle of Koenigstein.

Pasch managed to escape after two and a half years, but died shortly afterward, while Bötticher, who was allowed to carry on his experiments in the castle in the hope that he would be able to make a further supply of the powder, discovered a process for making a white porcelain superior to any then known. He thus became the inventor of the china for which Dresden has since become so famous. The Elector, on hearing this, gave him his freedom, hoping to profit by his inventions, but as a result of his long imprisonment Bötticher was taken ill and died in 1719 at the age of thirty-seven.

Meanwhile the mysterious Lascaris appears to have been travelling about Europe performing wonderful feats in transmutation in various cities. He is said to have enriched the Baron de Creux and replenished the coffers of the Landgrave of Hesse-Darmstadt with gold, while in Leipzig he showed a goldsmith of that city an ingot of solid gold which he declared he had made.

Stories of similar mysterious persons who wandered from country to country were common at this time, such as that of the fugitive who was given refuge by the Countess d'Erbach at the castle of Odenworld and who changed all her silver into gold as a parting gift. There is also the legend of the travelling alchemist who made gold in the presence of J. G. Joch at Dortmund in 1720, and the story of the nameless adept who, at the palace in Vienna, in July 1716 transmuted copper coins into silver by first making them red-hot, then sprinkling them with a powder and immersing them in an unknown liquid.

Of these tales, many of which are doubtless due to the imagination of their narrators, one of the most interesting is that related by Johann Friedrich Helvetius, the Dutch chemist. He relates his experience in his work entitled *The Brief of the Golden Calf; discovering the Rarest Miracle in Nature, how by the smallest portion of the Philosopher's Stone a great piece of common lead was totally transmuted into the purest transplendent gold at the Hague in 1666.*

Helvetius was a man of eminence in his time, holding the post of physician to the Prince of Orange. Writing on December 27, 1666, he says:

> In the afternoon came a stranger to my house at The Hague of honest gravity and serious authority, of a mean stature, a little

ALCHEMISTS OF THE XVIIth CENTURY

long face, black hair, not all curled, a beardless chin and about forty-four years (as I guess) of age, born in North Holland. After salutation he beseeched me on the first reverence to pardon his rude accesses, for he was a lover of the Pyrotechnian art. He asked me if I was a disbeliever as to the existence of a universal medicine which would cure all diseases unless the principal parts were perished or the predestined time of death had come?

I replied, "I never met with an adept or saw such a medicine, though I have frequently prayed for it."

"Surely, you are a learned physician?" I asked.

"No," said he, "I am a brass-founder and a lover of chemistry."

He then took from his bosom-pouch a neat ivory box and out of it three ponderous lumps of stone, each about the bigness of a walnut.

I greedily saw and handled for a quarter of an hour this noble substance, and having drawn from the owner many rare secrets of its admirable effects, I returned him this treasure of treasures, beseeching him to bestow a fragment of it upon me, though but the size of a coriander seed, but he refused.

He then asked me if I had a private chamber whose prospect was from the publick street, so I presently conducted him to my best furnished room, which he entered without wiping his shoes, which were full of snow and dirt.

He then asked for a piece of gold, and opening his doublet showed me five pieces of that precious metal which he wore upon a green riband.

I now earnestly craved a crumb of the stone, and at last he gave me a morsel as large as a rape seed.

"But," I said, "this scant portion will scarcely transmute four grains of lead!"

"Then," he said, "deliver it back," which I did in hopes of getting more, but he cutting off half with his nail said, "Even this is sufficient for thee. Even that will transmute half an ounce of lead."

So I gave him thanks and said I would try and reveal it to no one.

He then took his leave and said he would call again next morning at nine. I then confessed that while the mass of his medicine was in my hand I had secretly scraped off a bit with my nail which I projected on lead, but the whole flew away in fumes.

"Friend," said he, "thou art more dexterous in committing a theft than in applying medicine. Hadst thou wrapped up thy stolen prey in yellow wax it would have penetrated and transmuted the lead into gold."

The little man left, after promising to show Helvetius the manner of projection when he returned on the morrow. "But," says Helvetius,

> Elias never came again; so my wife, who was curious in the art whereof the worthy man had discoursed, teased me to make the experiment with the little spark of bounty he had left me. So I melted half an ounce of lead upon which my wife put in the said medicine. It hissed and bubbled, and in a quarter of an hour the mass of lead was transmuted into fine gold, at which we were exceedingly amazed. I took it to the goldsmith who judged it most excellent and willingly offered fifty florins for each ounce.

Such is the story recorded by Helvetius, who at least was convinced that he had carried out a real transmutation, but nothing more is known of the mysterious stranger whom he believed to be the possessor of the much-sought-for Stone.

CHAPTER XXI

BEN JONSON'S "ALCHEMIST" AND OTHER QUACKS

THE seventeenth century, which was destined to see the birth of scientific chemistry, saw alchemy gradually falling entirely into the hands of the charlatans, who found in it an easy means of making gold through the gullibility of the more ignorant. The mystery in which the art was enveloped gave these cunning knaves an advantage which they were not slow to use for their own purposes. They swarmed over Europe, and while the honest alchemists pursued their work the quacks, who added to their rogueries astrology and fortune-telling, brought contumely upon it.

The literature of the period bears evidence of how alchemy had become degraded. Butler satirized the art in his *Hudibras*, and Ben Jonson held the mirror to the foibles of the credulous rich in *The Alchemist*. This comedy, which was written in 1610, is evidently typical of the time, and tells the story of a London household, the master of which flies from the city in fear of the plague.

Jonson introduces us to Subtle, his butler, who in his master's absence, aided by two accomplices, Face and Dol Common, takes the opportunity of playing the professional quack and sets out to dupe the public. The three impart marvellous information to Drugger, a brainless tobacconist, and perform mysterious rites with Surly, a gamester, while they seek to ensnare a greedy knight, Sir Epicure Mammon, who wishes to add to his wealth by means of alchemy. Two Anabaptists named Tribulation and Ananias, who hope by means of the Philosopher's Stone to be able to sustain the Puritan religious system, also take part in the play.

ALCHEMY AND ALCHEMISTS

Mammon, who is in haste to make gold, takes Surly to the alchemist and tells him:

>   This night I'll change
> All that is metal in my house to gold :
> And, early in the morning, will I send
> To all the plumbers and the pewterers,
> And buy their tin and lead up ; and to Lothbury
> For all the copper.
>
> . . . . .
>
> But when you see th' effects of the Great Med'cine,
> Of which one part projected on a hundred
> Of Mercury, or Venus, or the Moon,
> Shall turn it to as many of the Sun ; [1]
> Nay, to a thousand, so *ad infinitum.*
>
> . . . . .
>
> The perfect ruby, which we call elixir,
> Not only can do that, but by its virtue,
> Can confer honour, love, respect, long life ;
> Give safety, valour, yea, and victory,
> To whom he will.   In eight and twenty days,
> I'll make an old man of fourscore, a child.

Surly, who is sceptical of the power of the alchemist, observes:

>   Alchemy is a pretty kind of game,
> Somewhat like tricks o' the cards, to cheat a man
> With charming.
>
> . . . . .
>
>   What else are all your terms,
> Whereon no one o' your writers 'grees with other ?
> Of your elixir, your *lac virginis*,
> Your stone, your med'cine and your chrysosperm,
> Your sal, your sulphur, and your mercury,
> Your oil of height, your tree of life, your blood,
> Your marchesite, your tutie, your magnesia,
> Your toad, your crow, your dragon, and your panther ;
> Your sun, your moon, your firmament, your adrop,
> Your lato, azoth, zernich, chibrit, heautarit,
> And then your red man, and your white woman,
> With all your broths, your menstrues, and materials.

---

[1] Gold.

SOME QUACK ALCHEMISTS

Later Subtle puts his accomplice, Face, through the following amusing examination of his knowledge in order to convince Ananias of his skill:

> Sirrah my varlet, stand you forth and speak to him
> Like a philosopher : answer ; i' the language.
> Name the vexations, and the martyrizations of metals
>   in the work.

Face thus recites:

>   Sir, putrefaction,
> Solution, ablution, sublimation,
> Cohobation, calcination, ceration, and
> Fixation.
>
> SUBTLE. And when comes vivication ?
> FACE. After mortification.
> SUBTLE. What's your *ultimum supplicium auri* ?
> FACE. Antimonium.
> SUBTLE. Your *lapis philosophicus* ?
> FACE.   'Tis a stone,
> And not a stone ; a spirit, a soul, and a body :
> Which if you do dissolve, it is dissolv'd ;
> If you coagulate, it is coagulated :
> If you make it to fly, it flieth.

At the close of the play the absent master of the house suddenly returns, but is afraid to enter, for they announce that it is infected by the plague; but in the end the butler confesses his masquerade and all is forgiven.

Of the type of charlatan satirized by Ben Jonson was Simon Forman, who claimed to be an alchemist, astrologer, and magician. He was born in 1552 and, after acquiring some knowledge at Oxford, came to London to practise physic. From time to time he was brought before the College of Physicians and fined for pretending to cure the sick, and was several times sent to prison. In 1594 he began to experiment in transmutation and telling fortunes, and he soon attracted many customers, mostly women of good position. William Lilly calls him "a very silly fellow, yet had wit enough to cheat ladies and other women by pretending skill in telling their fortunes, as to whether they should bury their husbands and what second husband they should have." He is thought to have been the original of Subtle in Ben Jonson's *Alchemist*. He left behind him a diary and voluminous manuscripts on astrological and alchemical subjects which are now in the Ashmolean Collection in the Bodleian Library at Oxford.

At the trial of those charged with the murder of Sir Thomas Overbury in 1615 it was discovered that a Mrs Turner, who was implicated in the crime, had constantly consulted Simon Forman on behalf of the Countess of Essex, who was seeking to obtain a divorce from her husband. He was asked to furnish philtres which would alienate the Earl from the Countess and draw toward her the love of the Earl of Somerset.

Ben Jonson refers to the fame of his love-philtres in *Epicene*, and in Richard Nicols's poetical account of the murder he is thus alluded to:

> Forman was, that fiend in human shape,
> That by his art did act the devil's ape.

Contemporary with Forman was Francis Anthony, who became famous for his 'Potable Gold,' which he claimed was the elixir capable of curing all diseases. His process was published in 1683 under the title *A Receit showing the Way to make the most Excellent Medicine called AURUM POTABILE*. The sale of his elixir was so successful that it was carried on by his son for some time after his death.

About 1680 an alchemist, who claimed to be in possession of the true Philosopher's Stone, took up his residence in Light's Court, near the King's Arms, close to St Giles' Church in London. According to a bill exhibited over his door, "He hath brought along with him the work of a Famous Philosopher which is the True Matter and Stone of Philosophers and Naturalists concerning Gold and Silver."

It further states that he was ready to show for a shilling

> some Sulphur and Mercury in their crudity and also in their marriage, and he did this to undeceive so many people and to

SOME QUACK ALCHEMISTS

hinder many Learned and Chymists from wasting their estates and consuming their lives to no purpose, to come to the knowledge of the great work to which they shall never attain, as long as they shall make use, as the most part do, of matters which are quite contrary to that end.

He continues:

By seeing this work and by working they may, if it please God, come to the knowledge of the True Philosophical Sulphur, Salt, and Mercury fixed by Nature, and see that possible which many believe Impossible, and they will afterwards employ their time upon a natural subject and common to all mankind which they shall never want and will return the Gentleman thanks for having shown them their errors and freed them from great expenses.

Another well-known quack-alchemist who flourished in London about this time was Moses Stringer, who claimed to have discovered an "elixir capable of renewing youth to the aged and prolonging life." In a letter addressed to the "Learned Dr Woodrofe, Master of Worcester College in Oxford," Stringer thus describes the properties of his wonderful discovery:

Since I had the honour of your instructions in the University concerning physick and chemistry, I have in a particular manner apply'd myself to the study of those sciences. I have considered the nature of Humane Bodies and consulted the history of the ancients tho' I can't give credit to what the Poets record of Æson; yet what Paracelsus reports concerning the force of medicines in Recovering Old Age affects me very much.

That learned Chymist made his first experiment upon a Hen, so very old that nobody would kill it, either out of sense of profit or good-nature. He mingled some of his medicine, which he called "Renovating Quintessence," with a quantity of Barly and gave it to the Hen fifteen days together. The effects were wonderful and the Hen recovered Youth and New Feathers, and what is still more surprising LAID EGGS and Hatcht chickens, as if she had lost a dozen years of her age.

An ancient woman that kept his house with the consequences of Old Age was upon the very margin of death. He gave her the same medicine fifteen days together as he had prescribed to his feathered patient and the success was the same. She recover'd her Health,

Youth, Hair, and Teeth again. Her Complexion lookt florid and vigorous and Nature exerted itself as it generally does in Young Women.

Stringer tells the doctor that, having reflected "upon these Cures with considerable cost and pains," he has discovered such a remedy to renew youth very much and help old age. He called it "Elixir Renovans, because it doth refresh and make young again and it was to be obtained only at his House, for fear of counterfeits, at a Guinea a Bottle, sealed with Three Eagles displayed."

Moses Stringer lived to a good old age at his house in Blackfriars, near Puddle Dock, but whether his longevity was due to his elixir or not there is no evidence forthcoming.

CHAPTER XXII
THE ROSICRUCIANS AND ALCHEMY

THE mystery of the Rosicrucians, that strange and secret fraternity of the Rosy Cross, which is supposed to have flourished between the fifteenth and seventeenth centuries, has never been satisfactorily solved. Did such a society ever exist or was its foundation merely a fable only believed in by some of the mystic alchemists of the latter period? The question still remains to be answered.

Study of the subject reveals that nothing was known of the fraternity until the early years of the seventeenth century; what appears to be the first indication of its existence is found in an anonymous pamphlet entitled *Fama Fraternitatis, or A Discovery of the most Laudable Order of the Rosy Cross*, said to have been circulated in manuscript about 1610 and afterward printed at Cassel in 1614. The publication of this pamphlet aroused great interest and excited much controversy in Germany at the time and also in the latter half of the century among both philosophers and scholars. Such men as Descartes and Leibnitz endeavoured without success to discover the existence of the fraternity or its members.

Some light was thrown on the mystery when it was alleged that the author of the pamphlet was a Lutheran named Johann Valentin Andreä, of Würtemberg. This theologian admitted that he was the author of another Rosicrucian work, printed in 1616, in which he stated that he intended to ridicule the mania of the times for occult marvels, and that he borrowed the idea of a fraternity from contemporary romances of chivalry and travel. He chose the Rosy Cross as the symbol of the order, firstly because it was an ancient symbol used in the occult sciences,

ALCHEMY AND ALCHEMISTS

and secondly because it occurred in the armorial bearings of his family. But although the authorship and the intention of the pamphlet were thus revealed, it was accepted seriously by many who professed belief in the fraternity as an ancient order.

The fraternity was supposed to possess immense power and its members to comprise the great alchemists of two centuries, the sure depositaries during that time of the ancient mysteries of the Hermetic Art. According to one account, the order was founded in 1408 by Christian Rosencreutz, a Dutchman, but in another the founder is stated to have been a German nobleman who lived between 1378 and 1484. The story as told in *Fama Fraternitatis* states that the brotherhood, or fraternity, was formed with the intention of bringing about a general reformation. The founder is alluded to as the

> most godly and illuminated Father, Our Brother C. R. C., a German, the Chief and original of the Fraternity, who hath much and long time laboured, who by reason of his poverty (although descended of noble parents) in the fifth year of his age was placed in a cloyster where he learned indifferently the Greek and Latin tongues and (upon his earnest desire and request) being yet in his growing years was associated to a brother P. A. L., who determined to go to the Holy Land. Although this brother died in Cyprus and so never came to Jerusalem, yet Our Brother C. R. C. did not return but shipped himself over and went to Damascus, intending from thence to go to Jerusalem; but by reason of the feebleness of his body he remained still there, and by his skill in physic he obtained much favour with the Turks and in the meantime he became acquainted with the Wise Men of Damascus in Arabia and beheld what great wonders they wrought and how Nature was discovered unto them.

This story further relates how C. R. C., who was but sixteen years of age, "yet of strong Dutch constitution," became very friendly and intimate with the Wise Men, who showed him secrets whereat he could not but mightily wonder. He learned the Arabic language, physick, and mathematics. After three years he went to Egypt, then came by the whole Mediterranean

THE ROSICRUCIANS AND ALCHEMY

Sea unto Fez, where they were most skilful in mathematics, physic, and magic. Here he was taught many secrets and remained two years, after which he sailed with many costly things to Spain. There he conferred with the learned, showing unto them the errors of their arts and prescribed their new Axiomata whereby all things might be fully restored.

    These Axiomata were to be a combination out of all faculties, sciences and arts and whole nature, which should only serve to the wise and learned for a rule; that also there might be a Society in Europe which might have gold, silver, and precious stones, sufficient to bestow them on Kings for their necessary uses and lawful purposes, with which [society] such as be governors might be brought up for to learn all that which God hath suffered man to know, and thereby to be enabled in all times of need to give their counsel unto those that seek it like the Heathen Oracles. . . .

    But C. R. C. was received with ridicule and after many painful travels returned to Germany and settled. Although he could have bragged with his art, but specially of the transmutation of metals, yet did he esteem more Heaven and Men the citizens thereof than all vain glory and pomp.

    After five years had passed the idea again came into the mind of R. C. of a Reformation, and together with three of his brethren, Brother G. V., Brother I. A., and Brother I. O., who had some knowledge of the arts. He did bind these three unto himself to be faithful, diligent and secret, as also to commit carefully to writing all that which he should direct and instruct them in the end, that those which were to come and through especial revelation should be received into this Fraternity.

    After this manner began the Fraternity of the Rosie Cross.

Eventually five other Brethren were added, making the fraternity nine in number, all bachelors and of vowed virginity. The objects of the brotherhood are set out as follows:

    I. That none of them should profess any other thing than to cure the sick, and that gratis.

    II. None of the posterity should be constrained to wear one certain kind of habit but therein to follow the custom of the country.

    III. That every year upon the day of C. they should meet

together at the house *Sancti Spiritus* (their headquarters built by C. R.) or write the cause of their absence.

IV. Every brother should look about for a worthy person who after his decease might succeed him.

V. The word R.C. should be their seal, mark and character.

VI. The Fraternity should remain secret one hundred years.

To these six articles the brethren bound themselves, and then separated into several countries.

"The first of the Fraternity which dyed, and that in England," continues the narrative, "was I. O., who was very expert. In England he is spoken of much, chiefly because he cured a young Earl of Norfolk of the leprosie."

Mention is made in *Fama Fraternitatis* of

> the ungodly and accursed gold-making which hath gotten so much the upper hand whereby under colour of it many renegades and roguish people do use great villainies and cozen and abuse the credit which is given them. Yea, nowadays men of discretion do hold the transmutation of metals to be the highest point of philosophy, but the true philosophers are of far another mind, esteeming little the making of gold which is but a paragon, for besides that they have a thousand better things.

The Brethren are earnestly admonished

> to cast away most of the worthless books of pseudo-chemists to whom it is a jest to apply the Most Holy Trinity to vain things or to deceive men with monstrous symbols and enigmas or to profit by the curiosity of the credulous.

From this it will be seen that the Rosicrucian doctrine only touched the spiritual side of alchemy, and most of the tenets were adapted from the works of Paracelsus, who was the first to develop the idea of the macrocosm and the microcosm being one. The fraternity acknowledged the idea of transmutation and called it "a great gift of God," but they condemned it as a purely physical process and allude to it as the "ungodly and accursed gold-making."

It is thought by some that prior to the end of the sixteenth

THE ROSICRUCIANS AND ALCHEMY

century there may have been an association of learned persons drawn from all classes, the members of which were engaged in healing, alchemy, philosophy, and good works, and that such a brotherhood gave Andreä the idea which he propounds in *Fama Fraternitatis*.

Michael Maier, the German alchemist, who was the first to transplant the Rosicrucian mystery into England, which he did when he came to see Robert Fludd, asserts that from "very ancient times philosophical colleges have existed among various nations for the study of medicine and natural secrets, and that the discoveries which they made were perpetuated from generation to generation by the initiation of new members."

The Rosicrucian legend was continued and fostered by a small band of men who looked upon it as a form of alchemical mysticism. In England Robert Fludd and Thomas Vaughan are regarded as its chief exponents, and afterward it was exploited by several pseudo-alchemists or quacks such as John Heydon and George Starkey.

So-called Rosicrucian societies arose in the eighteenth century claiming descent from the original fraternity, but they gradually died out. After 1750 the idea began to be propagated by the Freemasons, and in the system of high degrees in Scottish Freemasonry, especially in the Rosenkreuz degree, the symbols were retained with Masonic interpretations. Since 1866 colleges of a Masonic Rosicrucian Society have been formed in England, Scotland, and the United States, with the aim, it is said, of affording mutual aid and encouragement in working out the great problems of life and in searching out the secrets of Nature. Their objects, we are told, "are to facilitate the study of philosophy founded upon the Kabbala and the doctrines of Hermes Trismegistus, who was incalculated by the original Fratres Roseæ Crucis of Germany in 1450."

It would be interesting to know the results of such studies, and if they have yet succeeded in solving the mystery of the Emerald Table. But, in spite of its perpetuation, it cannot be said that

Rosicrucianism had any permanent influence on alchemy. The movement arose when the art, as practised until the seventeenth century, was on its decline, and its members were chiefly visionary theorists who did no practical work in laying the foundations of chemical science.

CHAPTER XXIII

MORE ALCHEMISTS OF THE SEVENTEENTH CENTURY

THE mystical aspect of alchemy was revived by several notable men in the early part of the seventeenth century. Some were mere visionaries who believed that the Hermetic Art was of spiritual attainment, its meanings veiled in allegory and illustrated by symbols, and that the making of gold was not its chief end and aim. By its study they claimed that man could be brought into closer communion with his Maker, and disease and suffering could be banished from the world. Others were adherents of the Rosicrucian movement, which at that time spread from Germany into Western Europe. Among the latter was Jakob Böehme (1575-1624), who claimed to have received a revelation in a wonderful vision in which the inmost secrets of Nature were revealed to him. He began life as a journeyman shoemaker, and travelled about the country, eventually settling at Gorlitz in Germany in 1598. Here he began to study alchemy, and after acquiring some knowledge of the art he employed its symbolic language in the elaboration of a system of mystical philosophy which he evolved. He described the Philosopher's Stone as the "Spirit of Christ" which must tincture the soul.

Another alchemist of this type was Michael Maier (1568-1622), who, after studying medicine and acting as secretary to the Emperor Rudolph, devoted himself to alchemy. He stated his belief that the ancient mythological legends of the gods and heroes of the Egyptians and Greeks concealed alchemical truths.

Among the Englishmen was Robert Fludd, who was born at Milgate House, Bearsted, Kent, in 1574. He graduated at

Oxford and received the degree of Doctor of Medicine in 1605, after which he travelled on the Continent, visiting France, Spain, Italy, and Germany. While abroad he became deeply interested in alchemy and returned with a considerable knowledge of the art. He then commenced practising as a physician in London and met with some success. He applied his mystical beliefs to medicine and claimed to influence the minds of his patients, thus aiding their cure. Fludd learned of the Rosicrucian movement from Michael Maier, who came to London to visit him. He was the author of numerous works, including the *Apologia Compendiaria Fraternitatem de Rosea Cruce suspicionis*, in which he defended the Rosicrucian doctrines against Libavius and contended that all true natural science was rooted in revelation. There is a monument and bust to Fludd's memory in Bearsted Church, where he was buried.

After him came Thomas Vaughan, the Welsh mystic alchemist and poet, who was born at Newton, in Brecknockshire, in 1622, and graduated as Bachelor of Arts at Jesus College, Oxford, where he afterward was made a Fellow. After fighting in the Royalist cause Vaughan settled in Oxford and became an ardent student of chemistry. He was a great admirer of Cornelius Agrippa, to whom in matters of philosophy he acknowledged that, "next to God, he owed all that he had." He held strong mystical views as regards alchemy and claimed to be "a philosopher of nature and no mere student of alchemy," which he stigmatized as a "torture of metals." He declared that the "true Philosopher's Stone was the Christian Philosopher's Stone, a stone often inculcated in scripture." "This," he says, "is the Stone of Fire of Ezekiel, the Stone with the Seven Eyes upon it in Zacharia and the White Stone with the New Name in the Revelation."

Although he published a translation of *Fama Fraternitatis*, he protests in the preface that he had "no relations with the Fraternity [of the Rosy Cross], neither did he desire their acquaintance." Vaughan wrote and published verse in English and in

A MYSTIC ALCHEMIST IN HIS LABORATORY
Viridarum Chymicum (1688)

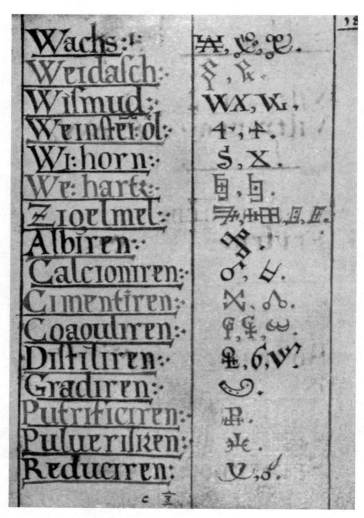

PAGE FROM A MANUSCRIPT ON ALCHEMY, DATED 1576,
BY CHRISTOPHER VON HIRSCHENBERG

This manuscript is said to have originally been in the library of the Emperor Rudolph II of Austria. It contains a list of chemical substances and their symbols.

By courtesy of Messrs Maggs Bros.

MORE XVIIth-CENTURY ALCHEMISTS

Latin, most of his works being written under the pseudonym of "Eugenius Philalethes." Among the books on alchemy attributed to him are *Lumen de Lumine* (1652), *Aula Lucis, or The House of Light* (1652), and *The Chymist's Key to Shut and to Open, or The True Doctrine of Corruption and Generation* (1657).

ALCHEMISTS AT WORK IN A LABORATORY
From *Theatrum Chymicum* (1693)

Vaughan died at the early age of forty-four, at Albury, owing, it is said, to the inhalation of poisonous fumes from some mercury with which he was experimenting.

Leaving the mystics for the more practical practitioners of the art, we find in Jean Baptiste van Helmont, who was born at Brussels in 1577, one of the most notable men of his time. With the intention of practising medicine, van Helmont studied for some years at the University of Louvain, where he eventually became Professor of Surgery, but, tiring of the routine of teaching, he resigned his post and, like Paracelsus, of whom he was a great admirer, he set out to travel. He gathered knowledge in all the countries he visited, and on his return resolved to apply himself seriously to the study of alchemy.

He was a man of marked religious temperament and, inspired by the work of Thomas à Kempis, gave his services freely to those who sought his advice. As an alchemist he firmly believed in the Philosopher's Stone and the Elixir of Life and claimed to have been successful in transmuting mercury into gold, but, like many of his predecessors, he did not know the composition of the 'projecting powder' he used. He states:

> He who first gave me the gold-making powder had likewise also at least as much of it as might be sufficient to change two hundred thousand pounds of gold. For he gave me perhaps half a grain of that powder, and nine ounces and three quarters of quicksilver were thereby transchanged. But that gold a strange man, being a friend of one evening's acquaintance, gave me.

Here again we have the oft-repeated story of the stranger who gave away the mysterious medium and then disappeared.

Van Helmont goes on to say regarding the 'powder of projection':

> I have divers times seen it and handled it with my hands, but it was of colour such as is saffron in its powder, yet weighty and shining like unto powdered glass.
> There was once given unto me one-fourth part of one grain; but I call a grain the six-hundredth part of one ounce. This quarter of a grain therefore being roaled up in paper, I projected upon eight ounces of Quicksilver made hot in a crucible, and straightway all the Quicksilver with a certain degree of noise stood still from flowing and being congealed settled like unto a yellow lump; but after poring it out, the bellows blowing, there were found eight ounces and a little less than eleven grains of the purest gold.

Whatever may be the explanation of this experiment, van Helmont was evidently convinced that he had effected a successful transmutation.

He did not believe in the third principle, which Valentine

called salt, and in his *Paradoxal Discourses* thus refers to the constitution of gold :

> Metals consist universally of a hot and cold sulphur. They are as male and female in respect to both of which, the more intimately they are united or naturally interwoven, the nearer the metals approach to the nature of gold.
> And from the difference and disparity of this union arises the distinction of all metals and minerals, that is in the proportions as the said sulphurs are more or less united in them.
> If metals be produced and consist by the union of these two, when then is there room for a third principle in metals which is vulgarly called salt?
> Ask Nature of what she makes gold and silver in the gold and silver mines, and she will answer, out of red and white arsenic, but she will tell thee withal that indeed gold and silver are made of the same.
> For the gold which is there in its vital place when it is wrought and made, is killed by the abundance of arsenic and afterwards made alive again and volatilized to bring forth other creatures as vegetables and animals and to give unto them being and life.

But, in spite of his belief in the delusions common in his time, van Helmont proved himself to be an original investigator and was practically the founder of pneumatic chemistry. He was the first to mention the different characters of gaseous substances, to give them the generic name of 'gas,' and to distinguish them from other vapours. Although he never succeeded in collecting them, he discovered and proved the presence of carbonic-acid gas, which he called 'gas sylvestre.' He showed how it was formed when charcoal was burned, and discovered its presence in the mineral waters of Spa. He classified gases as flammable and inflammable, and showed that air was not a modification of water.

Like Paracelsus, van Helmont was a firm believer in magnetism, and declared that it predominated everywhere and was an unknown property of a heavenly nature. He also recommended the employment of chemical substances in medicine, and was the first to use alum as a styptic in uterine hæmorrhage. He discounted the value of bleeding on the ground that it had a tendency to debilitate the patient. He was a pioneer in the analysis of urine, and, in 1655, was the first to make an investigation into its composition; as a result, he devised a method of examining it by weight and urged its importance in the diagnosis of disease.

Another practical chemist who flourished at this period was Johann Rudolf Glauber, who was born at Karlstadt, in Germany, in 1604. Like van Helmont, he began with the belief in the dreams of the alchemists, but in spite of this he carried out some valuable work in applied chemistry, and some of his processes have continued in use to the present day. He was the first to make sulphate of soda, which is still commonly known as Glauber's salt, and attributed to it remarkable medicinal virtues. This discovery was brought about by his analysis of the mineral spring at Neustadt, which he visited in order to take the waters and from which he derived great benefit. Before his visit the water from the spring was known as 'saltpetre water,' as it was believed to be impregnated with potassium nitrate, but Glauber proved it to contain sodium sulphate. He distilled ammonia from bones and demonstrated the art of making sal ammoniac by the addition of sea-salt. His ammonium sulphate is now used in the production of other ammonia salts and as a fertilizer. He also investigated pyroligneous acid, produced the chlorides of arsenic and zinc, and added considerably to the knowledge of the chemistry of wines and the distillation of spirits. He died at Amsterdam in 1668.

Mention should be made of Giambattista della Porta, a Neapolitan, for, although he achieved no great renown as an alchemist, he was a remarkable man in many ways and made discoveries in several branches of science. He was a profound student of physiognomy, and in one of his works makes some fantastic comparisons between the faces and features of human beings and those of birds and animals. He further attempts to show the similarity between certain animals and birds and

MORE XVIIth-CENTURY ALCHEMISTS

various parts of plants, and also suggests that some pieces of apparatus used by alchemists were so called from their resemblance to creatures. Thus, he compares the flower of the daisy to the human eye, and its root to the crayfish; the arum

THE ORIGIN OF THE PELICAN
Della Porta (1583)

THE MATRASS LIKENED
TO AN OSTRICH
Della Porta (1583)

lily to the uterus; the sweet-pea to the butterfly; the snapdragon to a calf's head; the flower of fennel to the crest of a peacock; larkspur to a crested lark; and the leaf of the thistle to a webbed foot. He likens the alembic of the alchemist to a prancing bear; the matrass to an ostrich with outstretched neck; the retort to a wild goose; and the pelican to the long-necked bird of that name.

His chief work, *De Distillationibus*, was printed in Rome in 1608, and contains many woodcuts depicting the apparatus used in distillation. He also wrote a treatise on *Natural Magic*, which passed through many editions, and performed numerous optical experiments, in the course of which he invented the camera obscura. He died in 1615.

CHAPTER XXIV
THE LAST OF THE ALCHEMISTS

THE story of James Price, who was among the last to believe in the alchemical doctrine of transmutation, is one of romantic and tragic interest. He was born in London in 1752, and his real name was James Higginbotham, but, to comply with the wish of a relative who left him a legacy, he became known as Price. He had a brilliant career at Oxford, and at the age of twenty-five took his degree as Master of Arts and was a gentleman commoner of Magdalen Hall. In 1778 the university, in recognition of his discoveries in natural science, especially in chemistry, conferred upon him the degree of Doctor of Medicine, and he was elected a Fellow of the Royal Society at the early age of twenty-nine.

Possessed of ample means, he took a country-house at Stoke, near Guildford, where he equipped a fine laboratory in which to continue his work in chemistry and with the aim of discovering a medium by means of which he could change certain metals into gold. After some years of labour he became convinced that at last he had found the long-sought-for secret, and invited a number of distinguished scientific men, including Lord Onslow and Lord Palmerston, the father of the famous statesman, to his house to witness his experiments in transmuting mercury into silver and gold. This, he said, he was able to carry out by means of a white powder which, he declared, was capable of converting fifty times its own weight of mercury into silver, and a red powder by means of which he was able to change sixty times its own weight of mercury into gold. The composition of these powders he kept secret, but he claimed that in seven successive trials he had mixed them in a crucible with mercury; at first

THE LAST OF THE ALCHEMISTS

four crucibles with weighed quantities of the white powder, and then three other crucibles with weighed quantities of the red powder, with the result that silver and gold appeared in the crucibles after they had been heated in the furnace. He had submitted the precious metals produced to assayers, who had pronounced them genuine.

Specimens of the gold were shown to George III, and Price published a pamphlet entitled *An Account of Some Experiments*, in which, although he repudiated the doctrine of the Philosopher's Stone, he claimed that he had "by laborious experiments discovered how to prepare these composite powders, which were the practical realizations of that long-sought marvel."

The pamphlet was translated into several languages, and it created considerable excitement in the scientific world. Some of the older Fellows of the Royal Society urged Price to reveal the secret of the preparation of the powders and pointed out the propriety of proving his discovery before the society of which he was a Fellow. Price, however, refused to comply with the request, and in a letter to a friend gives for his reason that he might have been deceived by the dealers who had sold the mercury to him, and that it might have contained gold. He was, however, further pressed by two leading Fellows of the Royal Society to repeat his experiments in their presence, but in reply to this he made the excuses that the powders were exhausted and that the expense of making more was too great for him to bear, while the labour involved had already affected his health.

The Royal Society now interfered officially, and the President, Sir Joseph Banks, and some of the Council insisted that, for the honour of the society, he must repeat his demonstration before its delegates and prove that his "statements were truthful and his experiments without fraud."

Under this pressure Price had no alternative but to accept the challenge, and at length consented to prepare within six weeks ten powders similar to those which he had used in a previous demonstration. Meanwhile, driven into a corner and knowing full well he would have great difficulty in deceiving the expert delegates of the society, he was racked with anxiety and at last became desperate. It is said that he made a rapid survey of the works of the German alchemists as a forlorn hope, trusting that he might chance on some successful method of transmutation in their writings, but without avail.

The day appointed in August 1783 arrived, and the three experts delegated by the Royal Society journeyed down to Price's house. Knowing his demonstration to be foredoomed to failure and that his career would thus be blasted, Price had prepared, a day or two previously, some concentrated cherry-laurel water, which he knew to be a powerful poison, and kept it in his pocket. He appears to have received his visitors calmly, and, after he had introduced them into his laboratory, in the course of making preparations for the demonstration and unseen by them, he quickly poured the cherry-laurel water into a glass and swallowed it. In a few moments, to the astonishment of the delegates, he collapsed and died in their presence. So ended the brilliant James Price at the age of thirty-one. Whether he was really an impostor with an overweening desire for notoriety or whether his brain had become deranged in the course of his studies will never be known.

The mystery of another alchemist, who lived in the village of Lilley, between Luton and Hitchin, in the early part of the nineteenth century, has never been solved. He was known as John Kellerman, and was said to have been the son of a certain John Kellerman of Prussian birth, a man of gigantic stature who, in order to avoid being pressed into the regiment of giants formed by Frederick the Great, fled to the West Indies and there married a creole.

Sir Richard Phillips, in his *Personal Tour through the United Kingdom* (1828), puts on record all that is known of this mysterious individual, who is said to have practised alchemy amid his rural surroundings in Hertfordshire. To quote Sir Richard's narrative:

THE LAST OF THE ALCHEMISTS

I learnt that he had been a man of fashion, and at one time largely concerned in adventures on the turf, but that for many years he had devoted himself to his present pursuits.

For some time past he had been inaccessible and invisible to the world; his house being shut and barricaded and the walls of his grounds protected with hurdles and with spring guns, so planted as to resist intrusion in every direction.

The house was in a horrid state of dilapidation.

A young man took charge of my card and went to the back part of the house to deliver it. He returned saying his master would be happy to see me.

Sir Richard describes the alchemist as being

a man of about six feet in height. On his head was a white night-cap, and his dress consisted of a long great-coat that had once been green, and a kind of jockey's waistcoat with three tiers of pockets.

His physiognomy was extraordinary. His complexion a deep sallow and his eyes large, black and rolling. His manner was polite.

The alchemist conducted his visitor into a very large parlour, and, having locked the door, put the key in his pocket. Sir Richard continues:

The room was a realization of a picture by Teniers. The floor was covered with retorts, crucibles, alembics, jars and bottles of various shapes, intermingled with old books piled upon each other, with a sufficient quantity of dust and cobwebs. Different shelves were filled in the same manner and on one side stood his bed.

In a corner, somewhat shaded from the light, were two heads with white and dark wigs on them. . . .

Having told him of the reports I had heard of his wonderful discoveries, he gave me a history of his studies, mentioning some men whom I happened to know in London who, he alleged, had assured him that they had made gold.

Having studied and examined the works of the ancient alchemists from the multitude, he had pursued their system under the influence of new lights and after suffering numerous disappointments he at length had made gold. He could make as much as he pleased even to the extent of paying off the National Debt. . . .

Expressing my satisfaction at his success, I asked him to show me some of the precious metal he had made. "Not so," he replied; "I will show it to no one. I made Lord Liverpool the offer, that if he would introduce me to the King I would show it to his Majesty, but Lord Liverpool insolently declined; I am therefore determined that the secret shall die with me.

"It is true I made a communication to the French Ambassador, Prince Polignac, and offered to go to France and transfer to the French Government the entire advantages of the discovery, but after deluding me I found it necessary to treat him with the same contempt as the others.

"Every Court in Europe well knows that I have made the discovery and they are all in a confederacy against me lest by giving it to one I should make that country master of the rest.

"The world, sir, is in my hands!" he exclaimed with great emotion.

I inquired why he shut himself up in so unusual a manner, and he said it was to protect himself against the Governments of Europe, who were determined to get possession of his secret by force.

"I have been fired at twice in one day, through that window," he exclaimed, "and three times attempted to be poisoned.

"They believed I had written a book containing my secrets, and to get possession of this book has been their object. To baffle them I burnt all that I had ever written, and I have so guarded the windows with spring-guns and have such a collection of combustibles in the range of bottles which stand at your elbow, that I could destroy a whole regiment of soldiers if sent against me." . . .

He lived entirely in that room, and locked up the other rooms every night with patent padlocks and sealed the keyholes.

On leaving the abode of the alchemist, I learned from a man who had lived with Kellerman for seven years that he was one of eight assistants that he kept for superintending his crucibles. The man protested that no gold had ever been made in the place, and on telling him that I had been with Kellerman in his laboratory, he was astounded and assured me that he carried a loaded pistol in every one of his six waistcoat pockets.

Kellerman had the reputation in the neighbourhood of being a magician, and at night few of the country-folk would go past his house, from the chimneys of which heavy clouds of smoke constantly belched as his furnaces were kept going.

THE LAST OF THE ALCHEMISTS

One day, however—when it is not stated—Kellerman suddenly and mysteriously disappeared. No smoke was to be seen issuing from the chimneys, and the house appeared to be deserted. Nothing more seems to be known of his departure, but it was rumoured that he had gone to Paris, where he was maintained by a nephew, one Captain William Roebuck, and presumably he there ended his days.

Another believer in alchemical doctrines was the eccentric Peter Woulfe who had chambers in Barnard's Inn when in London, but usually resided in Paris during the summer. It is said that he spent many years in searching for the Philosopher's Stone and the Elixir of Life without success.

Breakfasting usually at four o'clock in the morning, he would sometimes invite a few friends to the meal. Entrance to the chambers themselves could only be gained by means of a secret signal.

He was a Fellow of the Royal Society and contributed several papers of value to *Philosophical Transactions*, including one entitled *Experiments on some Mineral Substances*. He is said to have been the inventor of Woulfe's apparatus, which still bears his name. He died in Barnard's Inn in 1803.

CHAPTER XXV

THE DAWN OF SCIENTIFIC CHEMISTRY

ALTHOUGH the belief in alchemy had been on the decline since the end of the seventeenth century, it was not until the first half of the eighteenth century that the practicability of transmutation and the possibility of discovering the Elixir of Life came to be doubted. But even in the eighteenth century notable chemists such as Stahl and Boerhaave were not convinced that transmutation could not be carried out. There were still those who believed that the experiences recorded by van Helmont and Helvetius were true and that gold had thus been made.

In Germany the alchemical belief in the Elixir was ridiculed by Semler, who had received a sample of the 'salt of life' from Baron Hirschen, and, treating it as the Philosopher's Stone, was surprised to find some gold deposited in the crucible after heating it. Klaproth, however, a German chemist, on analysing the 'salt of life' found it to consist of a mixture of Glauber's salt and magnesium sulphate. The question then arose, where had the gold come from that Semler had found in his crucible? The mystery was solved when Semler's old servant confessed that, eager to humour his master, he had slipped a number of sheets of gold-leaf into the crucible with the 'salt of life'; thus Baron Hirschen's discovery was discredited.

The dawn of scientific or modern chemistry began to break about the middle of the seventeenth century, when earnest workers sought to free themselves from the old traditions and questioned theories which had prevailed in previous ages. Owing to the researches of these men the clouds which had so long enveloped the art of the alchemist began to roll away; his

THE DAWN OF SCIENTIFIC CHEMISTRY

processes were no longer wrapped in mystery, for the experimenters published their researches, and this facilitated an exchange of ideas, which was furthered by the growing habit, practised by men of similar interests, of meeting for discussion of their experiences and views.

These meetings, or societies, as they were afterward called, probably originated in Italy with the foundation of Porta's Secret Academy of Nobility in 1560, and, somewhat later, of the Academy of Lynxes in Rome by Prince Frederigo Cesi. This society, which consisted of four members, of whom Gallileo was one, met together to solve their problems. The Académie Française was founded in 1635 by Richelieu in Paris, and in 1665 Colbert established the Académie des Sciences, which published its first *Transactions* in 1699.

In 1645 several men interested in science, including Dr Wilkins and Dr Wallis, met in London to discuss philosophical questions and to report on the results of their experiments. The outbreak and troubles of the Civil War, and the consequent removal of the Court from London to Oxford in 1649, caused them to migrate to that city, but the change had its advantages, for here they came into contact with Robert Boyle, Dr Petty, and Seth Ward, all of whom were ardent students of the natural sciences. This little band of congenial souls first began to meet at each other's houses with the object of "discoursing and considering all philosophical enquiries and such as related thereto." They also frequently met in the rooms of Dr Wilkins in Wadham College, and they eventually formed what was first called the Philosophical Society of Oxford.

In 1658 Boyle removed to London, and the society formed in Oxford eventually became the Royal Society of London, which was first organized in 1660 and constituted by King Charles II on April 22, 1662, as "a body corporate and politic by the appellation of the President, Council, and Fellowship of the Royal Society of London for improving natural knowledge."

John Evelyn records in his *Diary* the first anniversary meeting on November 30, 1663. The society began to publish its *Philosophical Transactions* on March 6, 1664. From this date all the famous scientific discoveries made in this country have been brought before this parent of all other learned societies in Great Britain. The most distinguished men in science have been among its presidents, including Boyle himself, Sir Christopher Wren, Sir Isaac Newton, Sir Hans Sloane, Sir Humphry Davy, Lord Kelvin, and Lord Lister.

Robert Boyle, prominent among pioneers in chemistry, was the seventh son of Richard, Earl of Cork, and was born at Lismore, in Ireland, in 1627. He was educated at Eton, and, after travelling for some years on the Continent, took up his residence at Stalbridge, in Dorsetshire, where he equipped a laboratory so that he might carry on his studies in chemistry. In 1654 he went to reside in Oxford so that he might come into closer touch with the little circle of men who were deeply interested in science and philosophy. He took rooms in the High Street on the west side of University College, and there fitted up a laboratory in which he worked until he removed to London.

He tells us that "chymistery was his greatest delight," and, writing in 1649, he says, "Vulcan has so transported and bewitched me as to fancy my laboratory a kind of Elysium."

Wood, the Oxford historian, gives us a glimpse of the pioneers of science who at that time met and conferred together in the old university city on the Isis. He mentions

> Arthur Tillyard, an apothecary and great Royalist who in 1655 sold coffey publicly in his house against All Soules Coll. He was encouraged so to do by some Royallists now living in Oxon and by others who esteem'd themselves either *virtuosi* or wits, of which the chiefest number were of All S. Coll., as Peter Pett, Thom Millington (afterwards an eminent physician and knight), Tim Baldwin, Christop. Wren (afterwards a great architect), George Castle, Will Bull, etc. There were others also as Joh. Lamphire, a physician lately ejected from New Coll. who was sometimes the natural droll of the company; the two Wrens, sojourners in Oxon., Matthew and Thomas, sons of Dr Wren, Bishop of Ely, etc.

A LABORATORY IN 1747

Stills, serpent, aludel, pelican, retorts, and other pieces of apparatus are shown.
From an engraving

A LABORATORY
From an engraving (1747)

THE DAWN OF SCIENTIFIC CHEMISTRY

Boyle's laboratory became a centre for the students of the new chemistry, and through them many others began to take up the study of the fascinating science.

In this laboratory Boyle invented his pneumatical engine, or air-pump, and also wrote his famous treatise *The Skeptical Chymist*, which has now become a classic in the history of chemistry. In it he defines an element as a substance which cannot be decomposed, but which will enter into combination with other elements, giving compounds capable of decomposition into these original elements.

To Boyle we are indebted for the discovery of the dependence of the boiling-point of a liquid upon atmospheric pressure, and also the knowledge, which he illustrated, of the expansive power of freezing water. He further explained the action of the syphon and the effect of air on the vibration of a pendulum and on the propagation of sound. He made experiments upon the nature of flames and on the relation of air to combustion and on respiration. He was also the first to make hydrogen and to prepare acetone by the distillation of the acetates of lead and lime.

About 1665 he became deeply interested in the manufacture of phosphorus, a substance which had intrigued alchemists and chemists for many years. According to tradition, it was first prepared by an old alchemist called Brandt, who lived at Hamburg, and urine was its source of origin. Little, however, is known of Brandt, but he comes into the story in connexion with Johann Kunckel, the son of an alchemist attached to the Court of the Duke of Holstein.

Johann Kunckel was born in 1630, and, after gaining some experience in the laboratory of his father, he became chemist and apothecary to the Dukes Charles and Henry of Lauenberg. He graduated in chemistry at the University of Wittenberg, and was afterward placed in charge of the glass-works and laboratories of the Elector of Brandenburg at Berlin. This building was destroyed by fire, and some time after this catastrophe King

Charles XI of Sweden invited Kunckel to Stockholm, where he was eventually made a baron with the title of von Lowenstern.

His chief work, *Laboratorium Chymicum*, was not printed until after his death. In it he tells the story of an alchemist named Bauduin, who together with another alchemist named Fruben lived at Grossenhayn, in Saxony, about 1668. These men evolved a fantastic scheme to extract what they called *spiritus mundi*, or the Spirit of the World, by chemical processes. These consisted in combining the four elements, earth, air, fire, and water, in an alembic and distilling the quintessence of the whole. He states that they dissolved lime in nitric acid, evaporated it to dryness, and then exposed the residue to the air, letting it absorb humidity. This they again distilled and obtained what they called "humidity in a pure form," which they are said to have sold with considerable success.

One day Bauduin accidentally broke a vessel which contained some of the calcined nitrate of lime and observed that, after exposure to sunlight, it became luminous in the dark. Having to pay a visit to Dresden, he took a specimen with him. During his stay in the city he met Kunckel, who was at that time living there. The latter was very much interested and became anxious to learn how this luminous stone had been made, but Bauduin was loath to enlighten him. Kunckel, however, managed to obtain a small quantity of it and afterward began to experiment himself by treating chalk with nitric acid, and eventually succeeded in making some of the luminous body. In 1669 Kunckel, having occasion to go to Hamburg, took a specimen of his product with him and showed it to a friend. He, however, did not seem to be particularly surprised, and he told Kunckel that Brandt, an old alchemist in the city, had made and shown him something far more wonderful. Kunckel was very anxious to see this marvel, and with his friend hurried off to find Brandt, who, after some little inducement, produced a specimen of a remarkably luminous substance which he claimed to

THE DAWN OF SCIENTIFIC CHEMISTRY

have discovered in the course of his experiments in search for the Philosopher's Stone.

Kunckel at once wrote to a friend in Dresden telling him of the wonderful new 'phosphor,' as it was called, and mentioning Brandt. The friend, without informing Kunckel, immediately left for Hamburg and succeeded in buying for 200 thalers Brandt's secret for making the substance. Kunckel, who was determined not to be outwitted by his friend, managed to obtain some idea of Brandt's process, and, continuing to work on it, he at length succeeded in making some phosphor. He was the first to introduce it into medicine, and in his treatise *Chemische Ammerkungen*, printed in 1721, he recommends a preparation, called by him "luminous pills," which, he states, contains three-grain doses of his phosphorus product. The new substance aroused great interest in the scientific world, and soon became known throughout Europe.

About 1670 Robert Boyle, who had been travelling on the Continent, heard, while in Germany, of the discovery of the wonderful new chemical substance, and, becoming deeply interested in it, determined to learn all he could about its manufacture. On his return to London he obtained a specimen from Dresden which he handed to his head chemist, named Bilger, with instructions to experiment with it with the idea of making more of the substance.

About this time Boyle heard of a young German chemist, named Ambrose Godfrey Hanckwitz, who had come to London and was living in Chandos Street with his wife and family, and was anxious to obtain work. Boyle got in touch with him with a view to evolving a special process of manufacture, and Hanckwitz being a clever and industrious worker, they soon succeeded. Between them they built a spacious laboratory in Southampton Street, which at that time opened into Maiden Lane and extended downward toward the Strand. Here they began to manufacture phosphorus, or "Icy Noctiluca," as Boyle called the substance, and established one of the first chemists' shops in London. The product made in Maiden Lane, and commonly known as the 'English phosphorus,' soon had a large sale throughout Europe.

As a man Robert Boyle is said to have been kindly and courteous to all with whom he came into contact, and he appears to have had very few enemies. He never married, and remained to the end of his life a philosopher in the truest sense of the

DISTILLING PHOSPHORUS IN GODFREY'S LABORATORY

word and one who loved science for science's sake. He died in London in 1691 and was buried in the old church of St Martin's-in-the-Fields.

In his experiments on air Boyle helped to lay the foundations of pneumatic chemistry, for he recognized that the air was not a simple or elementary substance, but a heterogeneous body, or, as he called it, "an aggregate of effluviams from such differing bodies," and his observations were confirmed by another remarkable man, John Mayow, who, during his short life, made his mark on this period when the old alchemical theories were gradually being superseded by the accurate investigation of facts.

John Mayow was born in Cornwall in 1645, and went to Oxford with a view to studying medicine. He became a Fellow of All Souls College, and in 1667, at the age of twenty-two, took his degree. While in Oxford he became deeply interested in

THE DAWN OF SCIENTIFIC CHEMISTRY

experimental chemistry and devoted himself to the investigation of the composition of air, the result of which he published in his *Tractatus de Respiratione*. He showed from his experiments that air contains two gases, one of which supports combustion and the breathing of animals, while the other extinguishes fire. The one which he said was necessary for combustion and respiration he called "Spiritus Nitro-Aereus," or "fiery air," and the other, which was incapable of supporting combustion, he showed to be left after the removal of the "Spiritus."

Mayow proved that the air wherein a substance is burned or which an animal breathes diminishes in volume during the burning or breathing, and asserted that a substance which is being calcined lays hold of a particular constituent of the air, and not the air as a whole. Although he was thus so near the discovery of oxygen, his theories were not proved until the time of Priestley and Lavoisier, toward the end of the eighteenth century. He grasped the essential facts about the formation of acids and oxides and thus anticipated the results obtained by Lavoisier a century later. He left Oxford in 1675 to settle in Bath, where he practised as a physician; in 1678 he was elected a Fellow of the Royal Society and died a few months later, in 1679, at the early age of thirty-four. There is little doubt that had he lived he would have exerted a still greater influence on the development of chemistry.

No chemist before Mayow appears to have collected gases in flasks or vessels inverted over water and to have studied change of volume in the gas by observing the rise and fall of the water in the glass vessel—a discovery of great value and interest in itself. He was a man who lived fully a century before his time.

Another chemist who did much to influence the science of this period was Georg Ernst Stahl, who was born in 1660. To him is mainly due the doctrine of phlogiston, which was destined to affect the development of chemistry for more than half a century.

The word 'phlogiston,' by which Stahl named his theory, is derived from the Greek word *phlogistos*, meaning 'burnt.' The theory was founded on the idea that all combustible substances contain a common principle. He held that the phlogiston of a combustible thing escapes as the substance burns and, thus becoming apparent to the senses, is manifested as flame or fire. What remained after a substance had been burned was the original substance deprived of its phlogiston. Thus, to restore the phlogiston to the product of burning was to reform the combustible substance. Among these substances were included those metals which changed on heating, and it was taught that the earthy principle of a metal remained in the form of ash, or calx, as it was sometimes called, when the metal was calcined, or when it was deprived of its phlogiston. In other words, the metals were considered to be compounds consisting of a metallic calx, now called the oxide, combined with phlogiston.

This, in simple words, was the theory which displaced the belief of the earlier alchemists that metals were the results of the spiritual actions of the three principles sulphur, mercury, and salt. It was certainly an advance on the old doctrines which had held sway for centuries, and it marked the beginning of a new era in chemistry.

Although Hermann Boerhaave achieved more renown as a physician than as a chemist, he did much toward the development of the new science. He was born near Leyden, in Holland, in 1668, and became remarkable for his wide knowledge of medicine, chemistry, botany, and other branches of science. He studied medicine at the University of Harderwyk, and afterward became Professor of Physic at the University of Leyden, of which he was Rector in 1714. His reputation as a physician and teacher spread throughout Europe, and students came from many countries to attend his lectures. He was a profound believer in the value of chemistry to the art of medicine. His work entitled *Elementa Chemiæ*, printed in 1732, achieved a world-wide reputation and was translated into many languages,

A LABORATORY, SHOWING FURNACES, RETORTS, AND VARIOUS APPARATUS USED FOR DISTILLATION IN 1751
From an engraving

A LABORATORY, SHOWING FURNACES AND RETORTS USED FOR DISTILLATION IN 1751

From an engraving

THE DAWN OF SCIENTIFIC CHEMISTRY

being regarded as the most luminous chemical treatise of the time. Part of the work is devoted to the chemical principles relating to the elements and the decomposition of bodies, which he grouped under the heads of animal, vegetable, and fossil, and from this developed the science of organic and inorganic chemistry.

Boerhaave defined chemistry as "the art whereby sensible bodies contained in vessels or capable of being contained therein are so changed by means of certain instruments, and principally heat, that their several powers and virtues are thereby discovered with a view to the philosophy of medicine."

In 1718 he became Professor of Chemistry at Leyden, and in an address which he gave on September 21 in that

A LABORATORY
Kiessling, 1752

year he showed that the vagaries of the alchemists, the theories of fermentation and effervescence, the fixing and weighing of fire, the acid and alkali theory in physiology and medicine, and all the errors that the chemists of one period fell into had been corrected by the subsequent investigations of chemists themselves.

231

Apart from being a man of great learning and the most distinguished teacher of his time, he had a wide knowledge of languages. Burton, a contemporary, tells us that "the Latin he spoke in lectures or conversation was so clear that he could reveal the most abstruse points to the meanest capacities." He died at Leyden on September 23, 1738, in his seventieth year.

CHAPTER XXVI
ALCHEMY AND GOLD-MAKERS IN MODERN TIMES

THAT there are many credulous people who still believe in alchemy and the possibility of transmuting the baser metals into gold and silver is evident from accounts which have been published in recent times of men who have posed as adepts and succeeded in extracting thousands of pounds from the pockets of their dupes. So long as the greed for gold and the desire to amass wealth quickly exists there will be those who are ready to be deluded by the specious promises of cunning impostors.

That this should have been true in earlier times, when superstitious beliefs influenced all branches of science and emperors, kings, and princes retained alchemists in their pay in the hope of being able to replenish their depleted coffers, can to some extent be understood, but that the great advances of modern science can actually lead to the revival of the old beliefs in the medieval alchemist is more difficult of comprehension. Apparently recent discoveries have given the credulous renewed hopes of obtaining riches by some marvellous processes of which they are utterly ignorant, and "this," says H. C. Bolton, "is now fostered by devotion to esoteric studies."

From the middle of the last century societies have been formed in France devoted to the study of hermetic mysteries, the members of which claim to study the processes of the alchemists and compare them with the work of modern chemists, in the hope, no doubt, of finding a golden reward.

The modern disciple of Hermes still believes in the Emerald Table, and as late as 1854 Theodore Tiffereau sent to the

Académie des Sciences in Paris six treatises in which he claimed to have discovered a method of converting silver into gold.

Within recent years Edward C. Brice, of Chicago, claimed to have discovered a process for making gold from pure antimony. The United States Patent Office refused to allow him to protect his process until a test had been made by three assayers of the Mint who should conduct the trial under Brice's instructions. Brice agreed, and the test was arranged, but the assayers found that all commercial antimony contained traces of gold, and that Brice did not even recover the whole of this by his process.

Another American claimant was Edward Pinter, who announced in 1891 that he possessed the Philosopher's Stone which had the power of multiplying gold to three times its original bulk. In demonstrating this he took a sovereign and, melting it in a crucible, added a quantity of his mysterious powders to it. When it had cooled the gold in the crucible was found to weigh three times its former weight. An analysis of his powders showed that one contained a large percentage of precipitated gold and the other some calomel. But this did not deter him. He declared publicly that the gold, before being multiplied, must stand in a certain acid for eighteen days, and that during this time the fumes arising would be so noxious as to be dangerous to human life. These fumes, indeed, were so horrible that they drove all but himself from the room in which the operation was taking place. A jeweller took the bait, and Pinter, before he made the demonstration, asked him to deposit in an acid bath for eighteen days the 90,000 dollars that were to be multiplied. To this the jeweller agreed, and an empty house that could be shut up during the time was taken in Baltimore. Meanwhile Pinter was called out of town and had not returned when the eighteen days elapsed. The jeweller got anxious, and in the end had the house broken open, when, to his dismay, the 90,000 dollars which had been placed in the bath had disappeared.

ALCHEMY IN MODERN TIMES

About the same time yet another American, Dr Stephen H. Emmens, claimed to have discovered a body intermediate between gold and silver which was capable of being changed into gold. This he called argentaurum. He established a laboratory for the manufacture of gold from silver by his secret process, in which mechanical treatment played an important part. In April 1897 he sold to the United States Assay Office six ingots of an alloy of silver and gold for the gross sum of 954 dollars as proof of his success in transmutation, but since that time nothing seems to have been heard of him.

Europe also has not been without its alchemical revivalists, for at the end of 1930 one Heinrich Kurschaldgen, who had been an employee in a dye-works near Düsseldorf, in Germany, claimed that he had discovered a process for transmuting the baser elements into gold by splitting and recombining atoms. He first demonstrated his process before an interested lawyer, who was so convinced that he advanced a large sum of money for the prosecution of this profitable operation on a larger scale. According to one account of the demonstration made by Kurschaldgen in his laboratory, his apparatus consisted of a battery and several mysterious-looking boxes and bottles connected with one another by wires. One of the bottles was filled with a mixture of sand and tap-water. After some preliminary operations, no doubt designed to deceive his dupe, the radiation which was to split up and recombine the atoms was set to work, and in about twenty minutes the contents of one of the bottles, when shaken out, were seen to contain a few grains of what appeared to be gold.

A prominent business-man in Cologne was then initiated into the secret, and was so delighted by the result of another demonstration made in his presence by the obliging alchemist that he declared that he was ready to put £5000 into the business. He is also said to have been instrumental in forming a German-American syndicate to develop and exploit the great discovery, and this company at once contributed £2500 to carry on the work of manufacture. Kurschaldgen is said to have succeeded in obtaining about £10,000 from his credulous victims before the fraudulent nature of his manipulation of the sand and water used in his process was eventually proved. When at length the police were informed of his operations he was arrested and tried at Düsseldorf on the charge of defrauding a number of persons of £12,500. At the trial stories were told of his remarkable eloquence, by means of which and his command of a few technical phrases, together with the mysterious apparatus, he so easily persuaded wealthy people to finance his experiments. A technical witness stated that minute quantities of gold were actually found in the bottles of sand and water through which Kurschaldgen passed an electric current, but he pointed out that it was easy to introduce the particles into the bottles, which no doubt had been done. Another witness, who was a member of an important commercial firm, declared that he still believed in Kurschaldgen's ability to make gold and did not regret that he had let him have £750. However, Kurschaldgen was found guilty and sentenced to eighteen months' imprisonment.

The most successful quack alchemist of modern times, who even outrivals the famous Cætano, was Franz Tausend, whose exploits aroused great excitement and interest in Central Europe in 1929. This impostor, called by many of his compatriots the "Cagliostro of Aubing," was the son of a plumber and as a young man followed his father's calling at Aubing, near Munich. He became interested in the study of chemistry, and after reading all he could on the subject his ambition soared to extraordinary heights, for he aimed at evolving a new theory of atomic and molecular structure by means of which he hoped to astonish the world. He held that by skilful application of his ideas lead might be transformed into gold, and began to boast that he had actually performed this operation and could repeat it whenever he wished.

It was first necessary to find the capital to exploit his discovery,

ALCHEMY IN MODERN TIMES

and for this publicity was required. So Tausend advertised in the newspapers, and this brought him in touch with his first victim, a lady whom he persuaded to advance him £5000. On obtaining the money he resolved to remove the scene of his operations to the country and launch out on a large scale. He first bought a castle at Eppan, near the little town of Bolzano, in the South Tyrol, established himself with the title of Baron, and gave out that he was a doctor of chemistry. Having surrounded himself with an air of mystery, which served to spread his fame, he married a pretty restaurant waitress, whom he installed as the *châtelaine* of the castle.

He then set out to enlarge his sphere of operations and to attract the capitalists. He established a laboratory and travelled about Germany giving demonstrations which were so cunningly carried out that he deceived men of business and people of influence in different parts of the country. He got people of wealth to back him up, and companies were formed not only to make gold, but also to exploit Tausend's other reputed inventions, which covered a large field and included the cheapening of the production of aluminium, the preparation of morphine from common salt, the cure of foot-and-mouth disease, and a marvellous specific for arresting bleeding.

One success followed another. A "Tausend Research Company" was founded by thirty prominent persons, who are said to have included princes, generals, bank directors, and well-known industrialists. Money poured in on every side, and he deluded his victims by producing genuine gold now and again which he declared he had manufactured by his process.

Meanwhile his dupes were bound to secrecy, and were promised enormous profits, Tausend paying dazzling *interim* dividends when individual shareholders grew importunate. He protested that he was able to make gold from lead at the comparatively modest rate of 88 pounds, of the value of about £4488, a month, and that he expected to be soon in a position to turn out tons of the precious metal at a time. A well-known German general became interested in Tausend's discoveries and procured for him the equipment of larger laboratories to carry on his researches. To him Tausend revealed his supposed secret, and he became one of the thirty members of the "164 Society," so called after the key number of Tausend's "vibrational theory of atomic and molecular structure." This society was formed to develop the great secret. The other members were to divide 20 per cent. of the profits of the gold-making, which, however, were in no case to exceed £500,000.

There is little doubt that Tausend could have kept his profitable schemes going much longer had he not been seized with the ambition to play the part of a great country magnate. He began to spend money lavishly and bought another castle, where he and his wife entertained on an extravagant scale.

At length one of his chief victims, a Munich banker, became suspicious and informed the police, with the result that the supposed alchemist and his wife were arrested and lodged in gaol in Bolzano, whence they were transferred to Munich and committed for trial.

The hearing revealed some remarkable facts showing how easily many people of intelligence and education had been deceived and victimized by a clever rogue. Some of the witnesses suggested that Tausend was to some extent a victim of self-deception, and others apparently still believed in his ability to transmute lead into gold. One of these told the court that he himself had produced the precious metal while working alone according to Tausend's formulæ. Experts were called to prove that the metal, alleged to have been produced by Tausend, by means of which he had gulled his victims was not pure gold, but a familiar alloy with silver which was used in the manufacture of jewellery and was a regular article of commerce. Tausend is said to have secured between £50,000 and £75,000 by his frauds. The sums reported to the police totalled about

ALCHEMY IN MODERN TIMES

£37,000, but many of his victims preferred to suffer in silence rather than incur publicity.

The judge pronounced Tausend to be "a brazen and unscrupulous impostor whose guilt was attenuated only by the credulity of his dupes and the mischievous influence of his wife," and sentenced him to three years and eight months' imprisonment.

A few years ago an amusing account appeared in the newspapers of an Englishman who claimed to be able to convert lead into mercury and mercury into silver and gold. This pseudo-alchemist was a great believer in astrology, and declared that he could produce gold successfully only when "the heavens were balanced in equinox." According to the reporter who was present at a demonstration, the inventor wore a long yellow dust-coat and donned a Turkish fez. A huge cauldron was filled according to his directions and a fire lighted under it. The cauldron was then sealed, and all waited for something to happen. It did—a great explosion occurred, and the onlooker was nearly killed by a piece of the flying cauldron. There was no gold, and as, according to the operator, the experiment could be carried out only at the annual period, he was safe for another year.

Thus it will be seen that the lure of alchemy can still be as strong as it was centuries ago and that human nature has remained unaltered throughout the ages.

CHAPTER XXVII
WHAT WE OWE TO THE ALCHEMISTS

THERE has been a tendency in recent times to depreciate the practical work accomplished by the early alchemists, but Francis Bacon was right when he observed over three hundred years ago that "the search and endeavours to make gold have brought many useful inventions and instructive experiments to light." There is little doubt that the search for the Philosopher's Stone and the attempts to discover a method of transmuting metals were the cause of actual work with a variety of substances, and that as the result a considerable knowledge of materials used in applied chemistry was acquired.

We know that the Egyptians, over a thousand years before the Christian era, knew of alum, sulphur, sodium chloride, lead sulphate, red lead, copper oxyacetate, antimonium sulphide, and calamine, and must have known how to prepare some of them. Later we find that the Greeks had a knowledge of quicksilver, which is mentioned by Theophrastus about 300 B.C. They also used orpiment, the yellow sulphuret of arsenic, and zinc oxide, which they called *kadmia* or *pompholix* and the alchemists afterward named *lana philosophica*. Copper sulphate, which they knew first as *chalcanthum*, the Romans afterward called *atramentum sutorium*, and realgar and potassium carbonate, both mentioned by Dioscorides, were used before the fourth century of our era. Jábir-ibn-Hayyán, in whose works the word 'alkali' is first employed, was the earliest investigator to make nitric acid by distilling a mixture of saltpetre, copper vitriol, and alum. He called it *aqua dissolutiva*, and it was later known as *aqua fortis*. He also obtained sulphuric acid, for he mentions that

WHAT WE OWE TO THE ALCHEMISTS

when alum is strongly heated a spirit distils over which has a high solvent power.

The generic name *sal* included the vitriols, potash, soda, saltpetre, and alum; and the early alchemists rarely made a distinction between soda and potash, but they added the word *sal* to individual salts to distinguish one from another, as instanced in *sal petræ* and *sal maris*. The generic name *spiritus* was applied to volatile acids, as in the case of *spiritus salis* for hydrochloric acid, which is still commonly known as spirit of salt.

Jábir also described the preparation of mercuric oxide by calcining the metal mercury, and of sublimate (mercury perchloride) by heating a mixture of mercury, salt, alum, and saltpetre.

He is the first to mention white arsenic (arsenious acid), which he obtained by roasting realgar. He gave it the name by which it is still known to distinguish it from the red and yellow forms (realgar and orpiment).

Although silver nitrate was known to Jábir, it was not employed in medicine until the seventeenth century, when it was introduced by Angele Sala as *magisterium argenti* or 'crystal Diana.'

From the thirteenth to the seventeenth century many valuable discoveries were made by workers in alchemy. The preparation of alcohol is described as early as the twelfth century, and ammonium carbonate, called volatile alkaline salt, and ammoniated mercury were known in Lully's time. Calomel, originally known as *mercurius dulcis*, was known in the thirteenth century, and the discovery of potassium sulphate is said to have been made by Isaac of Holland in the fourteenth century, as well as by Croll, who mentions it later as the *specificum purgans* of Paracelsus.

Many new chemical substances are described for the first time in the works of Basil Valentine. He prepared hydrochloric acid by heating common salt and green vitriol, and knew that when it was mixed with *aqua fortis* it made what is now called *aqua regia* because it is capable of dissolving gold. This led to the production of *aurum potabile*, from which great results were anticipated.

Antimony and its compounds were carefully studied by Basil Valentine, who showed how to prepare antimony from the native sulphide (*stibium*) by fusing it with iron. He also gives formulæ for making antimony trichloride, commonly called butter of antimony, and for basic chloride of antimony, known as Algaroth's Powder. Definite details are given in his works regarding flowers of sulphur, and various sulphur compounds were used for the production of sulphur and other bodies. A particular variety of compounds, including zinc blende, galena, iron, and copper pyrites, was known under the name of *marcasitæ*. He prepared his green vitriol from pyrites, and brandy from fermented grape-juice. Finally, he examined the air of mines and suggested practical methods for determining whether it was respirable or not.

Zinc sulphate and ferric chloride were known in Basil Valentine's time, as was also the precipitation of metallic silver from a solution of its nitrate by means of copper or mercury.

To Paracelsus we owe the introduction of many chemical substances into medicine. He employed arsenic and antimony in the treatment of cancer and leprosy, and believed in the value of mercury as a curative agent. He made potassium arseniate by heating arsenic with potassium nitrate, and recognized the value of alcohol in extracting the properties of vegetable substances. He says in one of his works, "There is no better way of extracting the essence of roots and herbs than to cut them up as small as possible and boil them in strong wine in a closed vessel, separate them by straining and distil the liquid through an alembic."

Zinc chloride, first known as *oleum lapidis calaminaris*, was described by Glauber, who, as we have already seen, was also the first to prepare sodium sulphate. He discovered as well potassium permanganate, although its composition was not

WHAT WE OWE TO THE ALCHEMISTS

known until 1730. Phosphorus was discovered by the German alchemist Brandt, who prepared it from urine, and the preparation of tartarated antimony was first described by Mynsicht in 1631.

Many other substances that we owe to the alchemists might be enumerated, but sufficient have been mentioned to indicate how much we owe to the work they accomplished centuries ago.

After considering what has been done in the past we may ask, What of the future? Who shall venture to prophesy what triumphs may be achieved in structural chemistry or physics, or what great discoveries may be made, in the course of a few years? Our research workers were never more keen or better equipped than they are at present. Says Professor Soddy:

> Looking backward at the great things science has already accomplished, it can scarcely be doubted that one day we shall come to break down and build up elements in the laboratory as we now break down and build up compounds.
>
> Although we are as ignorant as ever how to act about transmutation, it cannot be denied that the knowledge recently gained constitutes a very great help towards a proper understanding of the problem and its ultimate accomplishment.

The discoveries of the Becquerel rays and the element radium have disclosed far-reaching possibilities, and, as Lord Kelvin has observed, the discovery of the properties of the latter body has opened our eyes to other discoveries never suspected or dreamed of. The extraction of radium from pitchblende would indeed appear to harmonize with the belief of the early alchemists of the growth of metals in the womb of Nature.

Science has shown that even the diamond can be transformed into graphite by a powerful electric current between carbon poles, and that both diamond and graphite can be indirectly converted into charcoal. Many years ago Moissan succeeded in making small diamonds by fusing charcoal in molten iron or silver and allowing it to crystallize from the solution under high pressure, and so the possibility of manufacturing precious stones was demonstrated. Sir William Ramsay's investigations into radium showed that after a lapse of time this substance gave off emanations that were capable of decomposing water into oxygen and hydrogen, an excess of the latter being produced, and further that two other rare gases, helium and neon, were present. Regarding his discovery he says, "We must regard the transformation of emanation into neon in presence of water as indisputably proved, and if a transmutation be defined as a transformation brought about at will by change of conditions, then this is the first case of transmutation of which conclusive evidence is put forward." [1]

As far back as 1904 the possibility of causing the atoms of ordinary elements to absorb energy was seriously discussed, and, if such hypotheses prove to be just, the transmutation of the elements no longer appears an idle dream. What the alchemist so long sought will have been discovered, and it is not beyond the bounds of possibility that it might lead to that other goal of the early philosophers, an 'elixir' capable of the prolongation of human life.

[1] *Journal of the Chemical Society*, vol. xciii (1908).

INDEX

Abu Mansur Muwaffah, 64
Abu'l-Quasin Muhammad ibn Ahmad-al-Iraqi, 66–67
Académie des Sciences, 223
Academy of Lynxes, 223
Æneas Baræus and transmutation, 43
Æthiops martial, 23
Agricola, Georg (Bauer), 172
Agrippa, Henry Cornelius, 163–166
Air, early beliefs concerning, 13
Albertus Groot (Albertus Magnus), rules of, for alchemists, 78–79
Alchemist, The, 197–199
Alchemist of Light's Court, the, 200–201
Alchemy, dawn of, 9–15; derivation of word, 10; and mysticism, 12; astrology and, 36–40; prohibited, 39–40, 140; in China, antiquity of, 50; early treatises on, 134–135; royalty and, 140–149
Alcohol, first prepared, 135
Alembic, the, 111
Algaroth's Powder, 242
'Alkali,' first use of word, 240
Allegorical figures, 128
Alloys, preparation of, noted in Greek papyri, 42
Alphabets, alchemical, 130–131
Aludel, the, 112
Andreä, Johann Valentin, 203–204
Anthony, Francis, 200
Apparatus, alchemical, 109–119; value of, in 1560, 118–119; symbols for, 126
Aqua dissolutiva, 240
Aquinas, Thomas, 80–82
Arab charlatan at Prague, 153–154
Arabian alchemists, 59–67
Aristotle's theory, 12–13
Arnaldus de Villa Nova, 82–83
Ashmole, Elias, 95, 101, 103

Assyrian knowledge of chemistry and the arts, 28–30
Astrology and alchemy, 36–40
Athanor, the, 113
Atomic theories, early, 14
Augustus of Saxony, Prince, 147
Aurum musivum, 24
Aurum Vellus, 136–137, 139
Avicenna, 64

Babylonian knowledge of chemistry and the arts, 28–30
Bacon, Roger, 91
Balloon, the, 113
Balneum Mariæ, origin of, 48
Baths, 116
Bauduin, 226
Becquerel rays, 243
Benter, David, 147
Bernard of Treves, 83–86
Bird, William, 99
Bloomfield, William, 100
Böehme, Jakob, 209
Boerhaave, Hermann, 230–232; definition of chemistry, 231
Borri, Giuseppe Francesco, 180–187
Bötticher, Johann Friedrich, 192–194
Boyle, Robert, 224–225, 227–228; discoveries of, 225; experiments on air, 228
Bragadino, Count Marco, 154–155
Brandt, 225, 226, 227
Breviary of Naturall Philosophy, A, 99
Brice, Edward C., 234
Brief of the Golden Calf, The, 194–196
Bronze used by the Egyptians, 19

Cætano, Domenico Manuel, 174–180
Calomel 241

245

ALCHEMY AND ALCHEMISTS

Calx jovis, 24
Canon's Yeoman's Tale, The, 103–106
Canterbury Tales, The, 17
Caput mortuum, 116
Cerussa, 23
Chalcanthum, 240
Charles II and his laboratory, 146
Charles VI of France, 148–149
Charles VII of France, 148, 149
Charles IX of France, 148
Charles XII of Sweden, 149
Charnock, Thomas, 99
Chaucer, Geoffrey, 102, 103–106
Chema, 11
Chemi, 10
Chemistry, alchemy and, 222–232
China, alchemy in, 49–53
Christian IV of Denmark, 149
Chymike, 10
Cinnabar, esteemed by the Chinese, 50
Cloche, the, 113
Compositiones ad tingenda, 134
Compound of Alchymie, The, 93–94
Concerning the Seven, 33
Condensers, 114–116
Confessio Amantis, 102–103
Copper, 19
Copper sulphate, 240
Cordova as centre of learning, 64
Corrosive sublimate, 20
Crocus martis, 23
Crocus solis, 18
Crucifix, the, 113
Cucurbit, the, 109, 112
Cupellation described by Jàbir, 62

DASTIN, JOHN, 99
De Lannoy, Cornelius, 142–144
De Rohan, Cardinal Prince, 149
Dee, John, 142, 151, 152
Democritus, 46, 47
Democritus of Abdera, 46–47
Deppel, 176
Diamonds, artificial, to make, 56
Doctrine of Democritus, The, 66
Dyeing known to the Egyptians, 28

EARTH, early beliefs concerning, 13
Egyptian goldsmiths, 27
Electrum, 18
Elementa Chemiæ, 230

Elementa chemica, elements, symbols for, 122
Elixir of Gold, 95
Elixir of Life, 74 ; in India, 5
Elixir Renovans, 202
Elizabeth, Queen, and alchemy, 142, 143–145
Ellis, Sir Thomas, 145–146
Emblems of operations, 127–129
Emerald Table of Hermes, The, 31–32
Emerald tablet, the, 31–35
Emeralds, artificial, to make, 56
Emmens, Stephen H., 235
English phosphorus, Boyle's, 225, 227–228
" Eugenius Philalethes," 211
Experiments on some Mineral Substances, 221

Fama Fraternitatis, 203, 204–207, 210
Faust, 106
Ferdinand III, Emperor, 148
Fire, early beliefs concerning, 13
Flamel, Nicholas, 87–90
Flores martias, 23
Fludd, Robert, 209–210
Forense, Francesco, 147
Forman, Simon, 199–200
Fraternity of the Rosy Cross, 203–207 ; its foundation and articles, 204–206
Frederick William I, of Prussia, 148
Frederick William II, of Prussia, 148
Fruben, 226
Furnaces, 116

Gallathea, 107
Geber—see Jàbir-ibn-Hayyán
Glauber, Johann Rudolf, 214
Glauber's salt, 214
Gnostics skilled in alchemy, 48
Goethe, description of an alchemist by, 106
Gold, 14, 17–18 ; early use of, 17, 26
Gold jewellery, early, 27
Gold-makers in modern times, 233–239
' Golden Drops ' of La Mothe, 23
Goulard's ' Extract of Saturn,' 24

INDEX

Gower, John, and alchemy, 102–103
Great Stone of the Philosophers, The, 160, 161–162
Greek alchemists, 41–48
Greek apparatus, 43, 109

HANCKWITZ, AMBROSE GODFREY, 227
Hayck, Thaddeus von, 151–152
Helmont, Jean Baptiste van, 211–214
Helvetius, Johann Friedrich, 194–196
Hermes and the Holy Writings, 10–11
"Hermes, his seal," 26
Hermes Trismegistus, 25
Hindu alchemy, 54–58
Hindu methods of making artificial precious stones, 56–57
Hollandus, Johann Isaac, 172
Honnauer, George, 155

IATRO-CHEMISTRY, 167–170
"Icy Noctiluca," 227
Iron, 22–23; worked by the Egyptians, 22
Ismail ibn Lhocine Toughrai, 66

JÁBIR-IBN-HAYYÁN (Geber), 59–62
Jade-wine Spring, 49
Jewish alchemists, 47–48
John the High Priest, 28
John of Preston, 100
John de Walden, 92
Jonson, Ben, and alchemy, 197, 199, 200

Kadmia, 240
Kellerman, John, 218–219; his laboratory, 219
Kelly, Edward, 142, 145, 151–152
Key of Solomon, The, 33
Key of Wisdom, The, 140
Keys of Providence, The, 66
Killing the metals, 55
Kimia, 10
'King Charles' Drops,' 146
Krohnemann, Colonel, 155–156
Kunckel, Johann, 225–227
Kurschaldgen, Heinrich, 235–236

Laboratorium Chymicum, 226
Laboratory and apparatus, alchemist's, 108–119; in India, 55–56
Lana philosophica, 240
Lapis infernalis, 18
Lascaris, 192–193, 194
Lead known to the Egyptians, 23
Leopold I, 148
Libavius, Andreas, 172–173
Libellus de Alchemia, 78–79
Licences to practice alchemy, granting of, 141–142
London, described by Agrippa, 164
Lucas van Leiden, 136
Lully, Raymond, 79–80
Lunar caustic, 18
Lutes, 117–118
Lyly, John, 107

Magisterium argenti, 241
'Magistery of Saturn,' 23
Maier, Michael, 209
Manuscripts, alchemical, 134–139
Mappæ Clavicula, 109, 134–135
Mary the Jewess, 47–48
Maslaman-al-Majriti, 64–66
Matrass, the, 112
Mayow, John, 228–229
Mercurial barometer, 21–22
Mercurius dulcis, 241
Mercury, 19–22
Metals and planets associated, 16–24
Metals known to Hindus, 55
Metals, symbols for, 123–124
Mineralibus, De, 31
Muehlendorf, Andreas von, 155
Müller, Johann Heinrich, 192
Mysticism and alchemy, 77–78

NĀGĀRJUNA, 55
Near East, alchemy in, 25–30
Norton, Thomas, 94–95; his Elixir of Gold, 95; his Elixir of Life, 95; his writings, 95–99

Occult Philosophy, 163
Old Book of Dr Synesius, Greek Abbot, The, 45
Oleum lapidis calaminaris, 242
Olympiodorus, 45–46

247

Open Entrance, The, 70
Operations, alchemical, symbols for, 125
Ordinall of Alkimie, 95, 96–98
Oxford, Boyle at, 224

PAPYRI, Græco-Egyptian, alchemical information in, 41–42
Paracelsus, 21, 166–172
Paykull, 149
Pelican, the, 112
Personal Tour through the United Kingdom, 218–220
Philosopher's egg, the, 114
Philosopher's Stone, 68–76 ; search for, 68–70 ; colour of, 71 ; preparation of, 71–74 ; virtues of, 74–76.
Philosophical Transactions, 221
Phlogiston theory, 229, 230
Phosphorus, discovery of, 225–228
Pills of Immortality, 51
Pills of the Moon, 18
Pinter, Edward, 234
Planets and metals associated, 16, 17
Poison detection, early method of, 57–58
Pompholix, 240
Porta, Giambattista della, 214–215
'Potable Gold,' 200
'Powder of Saturn,' 23
Prague, laboratory of Emperor Rudolph II at, 150–151 ; Arab charlatan at, 153–154
Prayers before alchemical experiments, 108–109
Precious stones artificially made, 30
Price, James, 216–218
Primal elements, early beliefs concerning, 12
Primum ens Melissæ, 75–76
Processes, alchemical, symbolic representation of, 128–130
Pseudo-alchemists, 78, 162
Purple of Cassius, 18, 30

QUICKSILVER, 20

RADIUM, 243 ; Sir William Ramsay's investigations of, 244
Red oxide of mercury, 20, 21, 66, 241

Retorts, the, 111–112
Revelation of the Hidden Key, The, 160
Rhazes, 62–64
Richthausen, 148, 157–158
Ripley, George, 92–94
Robert of Chester, 91
Rosencreutz, Christian, 204–205
Rosicrucians—*see* Fraternity of the Rosy Cross
Royal Society of London founded, 223
Royalty, alchemists and, 140–149
Rubies, artificial, to make, 57
Rudolph II, Emperor, 148, 150
Rupert, Prince, studies chemistry, 146–147

Sage's Step, The, 65–66
St Margaret's, Westminster, symbolic window, 201
Sal, 241
Sal jovis, 24
Salts of Mars, 22
Sapphires changed into diamonds, 57
Schwertzer, Sebald, 147–148
Scotta, Alessandro, 154
Secret Academy of Nobility, 223
Secretis Artis Naturæ, De, 91
Secretum Secretorum, 33
Seeds of metals, belief in, 16
Sendivogius, Michael, 190–192
Serpent, the, 113
Seton, Alexander, 188–190
Sextus Julius Africanus, 44–45
Sheba, Queen of, 26
Silver, 17, 18–19
Silver nitrate, 18, 241
Skeptical Chymist, The, 225
Soma rasa plant, 54
Specificum purgans, 241
Spirit of the World, the, 226
Spiritus, 241
'Spiritus Nitro-Aereus,' 229
Splendor Solis, 136, 139
Stahl, Georg Ernst, 229
Stalbridge, Boyle at, 224
Still, the, 111
Stoics, the, 48
Stringer, Moses, 201–202

INDEX

Substances, alchemical, symbols for, 124–125
Sulphur-mercury doctrine, 69
Symbols, alchemical, and their origin, 120–130; in Chinese alchemy, 53
Synesius, 45
Syriac manuscripts on alchemy, 66

Tan, meaning of, 49
Tantras, the, 54
Taoism and Chinese alchemy, 49
Tausend, Franz, 236–239
Theatrum Chemicum Britannicum, 95
Thölde, Johann, 159
Thurneisser, Leonhard, 156–157
Tiffereau, Theodore, 233–234
Time, symbols for, 126
Tin, 24
Tincture of the moon, 18
Topaz, artificial, to make, 57
Töpfer, Benedict, 152–153
Tractatus de Respiratione, 229
Tribicus, the, 43, 110

Trimosin, Solomon, 136–139; alchemical notes of, 139
Triumphal Chariot of Antimony, The, 160
Twins, 112, 119
Twelve Treatises of the Cosmopolitan, 191

VALENTINE, BASIL, 159, 162; his "eight gates," 128
Vanity of Sciences and Arts, 165–166
Vaughan, Thomas, 210–211
Verdigris, 19
Von Hohenheim, Philippus Theophrastus Bombastus (Paracelsus), 20, 166–172

WATER, early beliefs concerning, 13
William de Brumley, 92
Woulfe, Peter, 220

ZIEGLER, MARIE, 155
Zodiacal signs and alchemy, 36–37
Zosimus, 28, 43–44

A CATALOG OF SELECTED
DOVER BOOKS
IN ALL FIELDS OF INTEREST

A CATALOG OF SELECTED DOVER BOOKS IN ALL FIELDS OF INTEREST

CONCERNING THE SPIRITUAL IN ART, Wassily Kandinsky. Pioneering work by father of abstract art. Thoughts on color theory, nature of art. Analysis of earlier masters. 12 illustrations. 80pp. of text. 5⅜ x 8½. 23411-8

ANIMALS: 1,419 Copyright-Free Illustrations of Mammals, Birds, Fish, Insects, etc., Jim Harter (ed.). Clear wood engravings present, in extremely lifelike poses, over 1,000 species of animals. One of the most extensive pictorial sourcebooks of its kind. Captions. Index. 284pp. 9 x 12. 23766-4

CELTIC ART: The Methods of Construction, George Bain. Simple geometric techniques for making Celtic interlacements, spirals, Kells-type initials, animals, humans, etc. Over 500 illustrations. 160pp. 9 x 12. (Available in U.S. only.) 22923-8

AN ATLAS OF ANATOMY FOR ARTISTS, Fritz Schider. Most thorough reference work on art anatomy in the world. Hundreds of illustrations, including selections from works by Vesalius, Leonardo, Goya, Ingres, Michelangelo, others. 593 illustrations. 192pp. 7⅛ x 10¼. 20241-0

CELTIC HAND STROKE-BY-STROKE (Irish Half-Uncial from "The Book of Kells"): An Arthur Baker Calligraphy Manual, Arthur Baker. Complete guide to creating each letter of the alphabet in distinctive Celtic manner. Covers hand position, strokes, pens, inks, paper, more. Illustrated. 48pp. 8¼ x 11. 24336-2

EASY ORIGAMI, John Montroll. Charming collection of 32 projects (hat, cup, pelican, piano, swan, many more) specially designed for the novice origami hobbyist. Clearly illustrated easy-to-follow instructions insure that even beginning papercrafters will achieve successful results. 48pp. 8¼ x 11. 27298-2

THE COMPLETE BOOK OF BIRDHOUSE CONSTRUCTION FOR WOODWORKERS, Scott D. Campbell. Detailed instructions, illustrations, tables. Also data on bird habitat and instinct patterns. Bibliography. 3 tables. 63 illustrations in 15 figures. 48pp. 5¼ x 8½. 24407-5

BLOOMINGDALE'S ILLUSTRATED 1886 CATALOG: Fashions, Dry Goods and Housewares, Bloomingdale Brothers. Famed merchants' extremely rare catalog depicting about 1,700 products: clothing, housewares, firearms, dry goods, jewelry, more. Invaluable for dating, identifying vintage items. Also, copyright-free graphics for artists, designers. Co-published with Henry Ford Museum & Greenfield Village. 160pp. 8¼ x 11. 25780-0

HISTORIC COSTUME IN PICTURES, Braun & Schneider. Over 1,450 costumed figures in clearly detailed engravings–from dawn of civilization to end of 19th century. Captions. Many folk costumes. 256pp. 8⅜ x 11¾. 23150-X

CATALOG OF DOVER BOOKS

THE STORY OF THE TITANIC AS TOLD BY ITS SURVIVORS, Jack Winocour (ed.). What it was really like. Panic, despair, shocking inefficiency, and a little heroism. More thrilling than any fictional account. 26 illustrations. 320pp. 5⅜ x 8½.
20610-6

FAIRY AND FOLK TALES OF THE IRISH PEASANTRY, William Butler Yeats (ed.). Treasury of 64 tales from the twilight world of Celtic myth and legend: "The Soul Cages," "The Kildare Pooka," "King O'Toole and his Goose," many more. Introduction and Notes by W. B. Yeats. 352pp. 5⅜ x 8½.
26941-8

BUDDHIST MAHAYANA TEXTS, E. B. Cowell and others (eds.). Superb, accurate translations of basic documents in Mahayana Buddhism, highly important in history of religions. The Buddha-karita of Asvaghosha, Larger Sukhavativyuha, more. 448pp. 5⅜ x 8½.
25552-2

ONE TWO THREE . . . INFINITY: Facts and Speculations of Science, George Gamow. Great physicist's fascinating, readable overview of contemporary science: number theory, relativity, fourth dimension, entropy, genes, atomic structure, much more. 128 illustrations. Index. 352pp. 5⅜ x 8½.
25664-2

EXPERIMENTATION AND MEASUREMENT, W. J. Youden. Introductory manual explains laws of measurement in simple terms and offers tips for achieving accuracy and minimizing errors. Mathematics of measurement, use of instruments, experimenting with machines. 1994 edition. Foreword. Preface. Introduction. Epilogue. Selected Readings. Glossary. Index. Tables and figures. 128pp. 5⅜ x 8½.   40451-X

DALÍ ON MODERN ART: The Cuckolds of Antiquated Modern Art, Salvador Dalí. Influential painter skewers modern art and its practitioners. Outrageous evaluations of Picasso, Cézanne, Turner, more. 15 renderings of paintings discussed. 44 calligraphic decorations by Dalí. 96pp. 5⅜ x 8½. (Available in U.S. only.)
29220-7

ANTIQUE PLAYING CARDS: A Pictorial History, Henry René D'Allemagne. Over 900 elaborate, decorative images from rare playing cards (14th–20th centuries): Bacchus, death, dancing dogs, hunting scenes, royal coats of arms, players cheating, much more. 96pp. 9¼ x 12¼.
29265-7

MAKING FURNITURE MASTERPIECES: 30 Projects with Measured Drawings, Franklin H. Gottshall. Step-by-step instructions, illustrations for constructing handsome, useful pieces, among them a Sheraton desk, Chippendale chair, Spanish desk, Queen Anne table and a William and Mary dressing mirror. 224pp. 8⅛ x 11¼.
29338-6

THE FOSSIL BOOK: A Record of Prehistoric Life, Patricia V. Rich et al. Profusely illustrated definitive guide covers everything from single-celled organisms and dinosaurs to birds and mammals and the interplay between climate and man. Over 1,500 illustrations. 760pp. 7½ x 10⅛.
29371-8

Paperbound unless otherwise indicated. Available at your book dealer, online at **www.doverpublications.com**, or by writing to Dept. GI, Dover Publications, Inc., 31 East 2nd Street, Mineola, NY 11501. For current price information or for free catalogues (please indicate field of interest), write to Dover Publications or log on to **www.doverpublications.com** and see every Dover book in print. Dover publishes more than 500 books each year on science, elementary and advanced mathematics, biology, music, art, literary history, social sciences, and other areas.